EFFECT OF SULPHIDE ON ENHANCED BIOLOGICAL PHOSPHORUS REMOVAL

EFFECT OF SULPHIDE ON ENHANCED BIOLOGICAL PHOSPHORUS REMOVAL

DISSERTATION

Submitted in fulfillment of the requirements of

the Board for Doctorates of Delft University of Technology

and

of the Academic Board of the UNESCO-IHE

Institute for Water Education

For

the Degree of DOCTOR

to be defended in public on

30, January 2017, at 12:30 hours

in Delft, the Netherlands

by

Francisco Javier Rubio Rincon

Master of Science, UNESCO-IHE

born in Durango, Mexico

This dissertation has been approved by the
promotor(s): Prof. dr. D. Brdjanovic and
Prof.dr.ir M.C.M. van Loosdrecht

Composition of the doctoral committee:

Chairman	Rector Magnificus TU Delft
Vice-Chairman	Rector UNESCO-IHE
Prof. dr. D. Brdjanovic	UNESCO-IHE / TU Delft, promotor
Prod.dr.ir. M.C.M van Loosdrecht	TU Delft, promotor

Independent members:
Prof. dr.ir.J.J. Heijnen	TU Delft
Prof. dr.E.D van Hullebusch	UNESCO-IHE
Prof. dr.P.H. Nielsen	Aalborg University, Denmark
Prof.dr.ir.C.J.N. Buisman	Wageningen University, The Netherlands
Prof. dr.ir. J.B. van Lier	TU Delft, reserve member

This research was conducted under the auspices of the Graduate School for Socio-Economic and Natural Sciences of the Environment (SENSE)

CRC Press/Balkema is an imprint of the Taylor & Francis Group, an inform business

Published by:
CRC Press/Balkema
PO Box 11320, 2301 EH Leiden, The Netherlands
Pub.NL@taylorandfrancis.com
www.crcpress.com – www.taylorandfrancis.com
ISBN 978-1-138-03997-1

To Marielle

and

My family that is growing

Table of Contents

Abbreviations

ATP	Adenosine triphosphate
ATU	Allyl-N-thiourea
BOD	Biochemical oxygen demand
BPR	Biological phosphorus removal
C	Carbon
COD	Chemical oxygen demand
DGGE	Denaturing gradient gel electrophoresis
DO	Dissolved oxygen
DPAO	Denitrifying polyphosphate accumulating organism
EBPR	Enhanced biological phosphorus removal
FISH	Fluorescence *in situ* hybridization
GAO	Glycogen accumulating organism
GLY	Glycogen
HAc	Acetate
HPr	Propionate
HRT	Hydraulic retention time
ISS	Inorganic suspended solids
MAR	Microautoradiography
MLSS	Mixed liquor suspended solids
MLVSS	Mixed liquors volatile suspended solids
N	Nitrogen
NADH	Reduced form of nicotinamide adenine dinucleotide
OHO	Ordinary heterotrophic organism
OUR	Oxygen uptake rate
P	Phosphorus
PAO	Polyphosphate accumulating organism
PAO I	*Candidatus* Accumulibacter phosphatis clade I
PAO II	*Candidatus* Accumulibacter phosphatis clade II
PCR	Polymerase chain reaction
PHA	Poly-β-hydroxyalkanoates
PHB	Poly-β-hydroxybutyrate
PHV	Poly-β-hydroxyvalerate
PH2MV	Poly-β-hydroxy-2-methyl-valerate
PolyP	Polyphosphate
PolyS	Polysulphur
ppk	Polyphosphate kinase
S	Sulphur
SBR	Sequencing batch reactor
SRB	Sulphate reducing bacteria
SRT	Solids retention time
T	Temperature

TSS	Total suspended solids
VFA	Volatile fatty acids
VSS	Volatile suspended solids
WWTP	Wastewater treatment plant

Summary

The excess of phosphorus (P) and nitrogen (N) in water bodies can lead to the exponential growth of bacteria and algae. This process is commonly known as eutrophication and results in the depletion of oxygen which can cause the death of aquatic wildlife. From these nutrients (P&N), phosphorus is the key nutrient to avoid in order to prevent eutrophication, as a vast number of bacteria and algae can cover their nitrogen growth requirement from nitrogen gas fixation (Yeoman et al., 1988).

Most of the phosphorus discharged into water bodies comes from municipal wastewater systems (EEA, 2005). Thus, in order to control and prevent eutrophication of water bodies, an efficient phosphorus removal process at wastewater treatment plants is highly important (Yeoman et al., 1998). Whereas, the addition of chemicals such as iron can achieve good removal of phosphorus, its use considerably increases the operation and maintenance costs of treatment plants. Thus, whenever possible the biological phosphorus removal process is preferred (Metcalf & Eddy et al., 2003).

The enhanced biological phosphorus removal (EBPR) is a popular process due to high removal efficiency, low operational costs, and the possibility of phosphorus recovery (Metcalf & Eddy et al., 2003). The EBPR is carried by microorganisms capable to store phosphorus beyond their growth requirements. These poly-phosphate accumulating organisms (PAOs) proliferate in wastewater treatment plants (WWTP) by recirculating the sludge through anaerobic and anoxic/oxic conditions. The influent is directed to the anaerobic tank, and it contains high concentrations of volatile fatty acids (VFA) like acetate (HAc) and propionate (HPr; Barnard, 1975).

Nevertheless, the stability of the EBPR depends on different factors such as: temperature, pH, and the presence of toxic compounds(Satoh et al., 1994; Smolders et al., 1994b; Lopez-Vazquez et al., 2009). While extensive studies have researched the effects of temperature and pH on EBPR systems, little is known about the effects of different toxic compounds on EBPR. For example, sulphide has shown to inhibit different microbial activities in the WWTP (Koster et al., 1986), but the knowledge about its effects on EBPR is limited.

Sulphide is the end product of the sulphate reduction process. Sulphate reduction is an anaerobic process, which occurs in wastewater with a high content of sulphate. Sulphate rich wastewater can be generated via (i) infiltration of saline water into the sewage(van den Brand et al., 2015), (ii) discharge of industrial effluents (Sears et al., 2004), (iii) use of sulphate based chemicals during drinking water treatment (Bratby, 2016) and (iv) use of saline water as secondary quality water for toilet flushing (Lee et al., 1997). Sulphate can be reduced to sulphide during the conveyance of wastewater (sewage) or during anaerobic conditions at wastewater treatment plants.

Whereas the sulphide generated in the sewage can cause a shock effect on EBPR, the continuously exposure to sulphide can cause the acclimatization and adaptation of the biomass. However, the last would be possible only in case sulphate reducing bacteria (SRB) can proliferate in the WWTP and continuously reduce sulphate into sulphide in the anaerobic tank of the WWTP. Therefore, it is expected that the EBPR will be differently affected by sulphide generated in the sewage as opposed to sulphide generated in the anaerobic tanks of the WWTP.

This study focuses on the effects of sulphide on the different stages of PAO (anaerobic, anoxic, and oxic). This study differentiates between the point of formation of sulphide (sewage or WWTP) and its effect on EBPR. To do so experiments were designed and carried out with a PAO enriched biomass exposed, during short-term (hours) or long-term (weeks), to a range of sulphide concentrations.

Whereas, it is certain that sulphide can be generated in the sewage, it is unclear whether the exposure to oxygen and nitrate caused due to the recirculation of sludge through anaerobic and anoxic/oxic conditions would hinder the activity of sulphate reducing bacteria (SRB) in WWTP. Thus, in chapter two the residual effect of oxygen, nitrate, and nitrite on the sulphate reduction process was analysed with three electron donors (acetate, propionate, and lactate). In none of the cases sulphate reducing bacteria were irreversible inhibited, suggesting that sulphate reducers can proliferate at WWTP with a sludge recirculation through anaerobic, anoxic, and oxic conditions. The sulphate reducing process was most severely inhibited by exposure of the biomass to oxygen. While the sulphate reducing bacteria that uses acetate as electron donor, were the most inhibited SRB.

In chapter three, the direct and reversible effects of sulphide on the metabolism of PAO were assessed. To do so, different concentrations of sulphide were added to a PAO enriched biomass, which was not acclimatized to sulphide. It was observed that sulphide affected both the anaerobic and more severely, the aerobic metabolism of PAO. The effect of sulphide on the metabolism of PAO was partially reversible up to 22 mg H_2S-S/L. Nevertheless, a higher sulphide concentration (36 mg H_2S-S/L) was lethal to PAO, which resulted in aerobic phosphorus release. Sulphide affected most severely the growth of PAO as it was not possible to observe any ammonia consumption (which is normally associated with microbial growth) at any of the sulphide concentrations assessed.

At a later stage, the same biomass used for the experiments performed in chapter three was exposed to sulphide for a long-term period (months; chapter four). Up to 20 mg S/L, the effects of sulphide on the metabolism of PAO were similar to those gathered in chapter three, which resulted in the presence of 5 mg PO_4-P/L in the effluent. In order to prevent phosphorus in the effluent, based on the results gathered in chapter three, the aerobic phase was extended from 4h to 5h. The extended aerobic phase allowed the complete removal of phosphorus up to 100 mg S/L. At 100 mg S/L, the settleability of sludge decreased, which resulted in an increase of suspended solids in the effluent. Further microbial characterization showed that up to 65% of the biomass were filamentous bacteria (*Thiothrix caldifontis*). Moreover, through activity tests and mass balances, it was suggested that *Thiothrix caldifontis* is capable to anaerobically store carbon (mainly as PHA) and aerobically store poly-sulphur and poly-phosphate.

PAOs generate energy from the oxidation of PHA for the storage of poly-phosphate. It has been suggested that both oxygen and nitrate can be used for PHA oxidation. As the effect of sulphide might differ depending on the electron acceptor available (oxygen or nitrate), a similar study as the one performed in chapter three was attempted in chapter five but on denitrifying poly-phosphate accumulating organisms (DPAO). However, the activity tests did not show any anoxic phosphorus removal, hence the effect of sulphide on the anoxic phosphorus uptake could not be carried out. Instead, in chapter five the denitrification capacities of a PAO I and PAO I-GAO culture were analysed and compared. Based on the denitrifying activity of the cultures, it is suggested that PAO I prefers the use of nitrite over nitrate. On the other hand, it is suggested that nitrite can be generated by side-communities such as glycogen accumulating organism (GAO).

In order to asses if the lack of anoxic phosphorus uptake (using nitrate) as observed in chapter five, could be caused by an enzymatic response to different operational conditions, the bioreactor with the PAO culture used in chapter five was operated for a long term period under anaerobic-anoxic-oxic conditions varying (i) the synthetic media used (ii)the solids retention time (SRT) (iii) P/COD ratio fed, (iv) the nitrate dose, and (v) the aerobic SRT. Despite these attempts, under none of the operational conditions a considerable anoxic phosphorus uptake was observed. Moreover, during an activity test it was observed that phosphorus was anoxically released, which indicates that PAO I were not capable to oxidize PHA using nitrate to generate energy.

This research suggests that sulphate reducing bacteria can proliferate in WWTP, as they are reversibly inhibited by the recirculation of sludge through anaerobic-anoxic-oxic conditions. The research enhances the understanding of the effect of sulphide on the anaerobic-oxic metabolism of PAO. It suggests that the filamentous bacteria *Thiothrix caldifontis* could play an important role in the biological removal of phosphorus. It questions the ability of PAO to generate energy from nitrate respiration and its use for the anoxic phosphorus uptake. Thus, the results obtained in this research can be used to understand the stability of the EBPR process under anaerobic-anoxic-oxic conditions, especially when exposed to the presence of sulphide.

Samenvatting

Overmatige aanwezigheid van fosfor en stikstof in waterlichamen kan leiden tot exponentiële groei van bacteriën en algen. Dit proces is algemeen bekend als eutrofiëring en resulteert in zuurstofdepletie waardoor aquatisch leven kan afsterven. Aangezien een aanzienlijk aantal bacteriën en algen in staat is, de voor groei vereiste stikstof uit stikstoffixatie te verkrijgen, is fosfor limitatie in het bijzonder van belang om eutrofiëring te voorkomen (Yeoman et al., 1988).

De lozing van fosfor in waterlichamen vindt grotendeels plaats via de gemeentelijke afvalwaterzuiveringssystemen (EEA,2005). Derhalve is, ter beheersing en voorkoming van eutrofiëring, een efficiënt werkend fosforverwijderingsproces bij afvalwaterzuiveringsinstallaties (AWZI) van groot belang (Yeoman et al., 1998). Alhoewel toevoegingen van chemicaliën zoals ijzer het ook mogelijk maken om fosfor te verwijderen, leidt het gebruik van chemicaliën tot stijging van de zowel operationele als onderhoudskosten van de AWZI. Daarom gaat de voorkeur naar het biologische fosfaat verwijderings proces (Metcalf & Eddy et al., 2003).

Het EBPR (Enhanced Biological Phosphorous removal) proces is door het hoge verwijderingspercentage, de beperkte operationele kosten én de mogelijkheid tot fosforterugwinning een veelvuldig toegepast proces (Metcalf & Eddy et al., 2003). Het EBPR proces wordt gerealiseerd door micro-organismen die in staat zijn om fosfor op te nemen in hoeveelheden die hoger zijn dan de fosfor behoefte voor groei. Deze polyfosfaat accumulerende organismen (PAOs) groeien in afvalwaterzuiveringinstallaties (AWZI) door slib te circuleren in afwisselend anaerobe en anoxische/oxische condities.Het met hoge concentraties vluchtige vetzuren (Volatile Fatty Accids; VFA) zoals acetaat (HAc) en propionaat (HPr) wordt gevoed aan de anaerobe tank (Barnard, 1975).

Desondanks is de stabiliteit van de EBPR afhankelijk van verschillende factoren waaronder; temperatuur, pH en de aanwezigheid van toxische stoffen (Satoh et al., 1994; Smolders et al., 1994; Lopez-Vazquez et al., 2009). Alhoewel de effecten van temperatuur en pH op de EBPR systemen uitgebreid onderzocht zijn, is de kennis ten aanzien van de effecten van verschillende toxische stoffen op de EBPR beperkt. Sulfide bijvoorbeeld, laat een afremming van verscheidene microbiële activiteiten in de AWZI zien (Koster et al., 1986). De kennis ten aanzien van het effect van sulfide op de EBPR is echter gelimiteerd.

Sulfide is het eindproduct van het sulfaatreductieproces. Sulfaatreductie is een anaeroob proces wat zich voordoet in afvalwater met een hoge sulfaatconcentratie. Sulfaatrijk afvalwater kan gegenereerd worden door (i) zoutwater infiltratie in het riool (van den Brand et al., 2015), (ii) lozing van industrieel afvalwater (Sears et al., 2004), (iii) gebruik van sulfaat houdende chemicaliën tijdens drinkwaterbehandeling (Bratby, 2016) en (iv) gebruik van zoutwater als secondair water voor toiletspoelingen (Lee et al., 1997). Sulfaat kan zowel tijdens het afvalwater transport (riolering) of tijdens de anaerobe conditie in de AWZI tot sulfide gereduceerd worden.

Terwijl de in de riolering gegenereerde sulfide een shock effect op het EBPR proces kan veroorzaken, kan voortdurende blootstelling aan sulfide leiden tot acclimatisatie en adaptatie van de biomassa. Dit is echter alleen mogelijk, indien sulfaat reducerende bacteriën (SRB) zich in de AWZI vermenigvuldigen en de SRB de sulfaat in de anaerobe tank constant tot sulfide reduceren.

Derhalve is de verwachting dat de beïnvloeding van riolering gegenereerde sulfide op EBPR verschilt ten opzichte van in de anaerobe tank gegenereerde sulfide.Deze studie richt zich op de effecten van sulfide op het metabolisme van PAO in de verschillende stadia (anaeroob, anoxisch en oxisch) en maakt hierbij onderscheidt tussen het punt van sulfide-formatie (riolering of AWZI) en het effect hiervan op EBPR.

Om dit te bewerkstelligen zijn experimenten ontworpen en uitgevoerd waarin met PAO verrijkte biomassa gedurende zowel korte (uren) en lange termijn (weken) blootgesteld werd aan verscheidene sulfideconcentraties.Alhoewel vast staat dat sulfide in de riolering gegenereerd kan worden is het onduidelijk of blootstelling aan zuurstof en nitraat, veroorzaakt door recirculatie van slib door anaerobe en anoxische/oxische condities de activiteit van sulfaat reducerende bacteriën (SRB) in de AWZI verstoort.

Derhalve wordt in hoofdstuk 2 het residuele effect van zuurstof, nitraat en nitriet op het reductieproces geanalyseerd door middel van 3 elektronendonoren (acetaat, propionaat en lactaat). In geen van de gevallen werden de sulfaat reducerende bacteriën onherstelbaar geïnhibeerd. Dit suggereert dat SRB in de AWZI kunnen prolifereren gedurende slib recirculatie door anaerobe, anoxische en oxische condities. Het sulfaat reduceringsproces werd het ernstigst belemmerd door de biomassa aan zuurstof bloot te stellen, terwijl is gebleken dat de sulfaat reducerende bacteriën welke acetaat als elektrondonor gebruiken, de meest belemmerde SRB zijn.

In hoofdstuk 3 worden de directe en reversibele effecten van sulfide op het PAO metabolisme onderzocht.. Hiervoor zijn verschillende sulfideconcentraties aan een PAO biomassa cultuur toegevoegd, die niet geacclimatiseerd was aan sulfide. Hiermee werd vastgesteld dat sulfide, op zowel de anaerobe als in sterkere mate op het aerobe PAO metabolisme effect heeft. Het effect van sulfide op het PAO metabolisme was deels reversibel tot 22 mg H_2S-S/L. Desondanks is gebleken dat een hogere sulfide concentratie (36 mg H_2S-S/L) PAO fataal werd en resulteerde in aerobe fosfor afgifte. Het effect van sulfide was het voornaamst op de groei van PAO, aangezien ammonium consumptie, wat geassocieerd wordt met microbiële groei, in geen van de getoetste sulfide concentraties waargenomen werd.

Met biomassa cultuur die ook gebruikt is voor de experimenten in hoofdstuk 3 werd in een later stadium een lange termijn (maanden; hoofdstuk 4) experiment uitgevoerd waarin de biomassa blootgesteld werd aan sulfide. De effecten van sulfide op het PAO metabolisme waren tot 20 mg S/L, vergelijkbaar met de gemeten effecten in hoofdstuk 3, wat resulteerde in de aanwezigheid van 5 mg PO4-P/L in de effluent. Gebaseerd op de resultaten zoals beschreven in hoofdstuk 3 werd, om fosfor in de effluent te voorkomen, de aerobe fase 4 naar 5 uur verlengt.

De verlengde aerobe fase maakte het mogelijk om volledige fosfor verwijdering te realiseren tot sulfide concentraties van 100 mg S/L. Bij 100 mg S/L vermindert de slibbezinking met als gevolg een verhoogde concentratie zwevende droogstof in het effluent. Aanvullende microbiële karakterisering laat zien dat tot 65% van de biomassa bestaat uit filamenteuze bacteriën (*Thiotrix caldifontis).* Verder suggereren de activiteittesten en massabalansen dat *Thiotrix caldifontis* in staat is koolstof (hoofdzakelijk als PHA) onder anaerobe condities en polysulfaat en poly-fosfaat onder aerobe condities op te slaan.

PAOs genereren energie door middel van PHA oxidatie voor de opslag van poly-fosfaat. Het idee bestaat dat zowel zuurstof als nitraat voor PHA oxidatie gebruikt kunnen worden. Aangezien het effect van sulfide op het metabolisme mogelijkerwijs verschilt met verschillende beschikbare elektronenacceptoren (zuurstof of nitraat), is er in hoofdstuk 5 getracht een onderzoek uit te voeren, dat vergelijkbaar is met het onderzoek in hoofdstuk 3 op denitrificerende poly-fosfaat accumulerende organismen (DPAO).

Echter toonden de uitgevoerde activiteittesten geen anoxische fosfaat verwijdering aan waardoor het effect van sulfide op de de anoxische fosfaat opname niet meetbaar was.

Als alternatief werden de denitrificatie capaciteiten van PAO I en PAO I-GAO culturen geanalyseerd en vergeleken. Gebaseerd op de denitrificerende activiteiten van de betreffende bacterieculturen, wordt verondersteld dat PAO I het gebruik van nitriet boven nitraat prefereert. Verder wordt verondersteld dat nitriet door side-communities zoals glycogeen accumulerende organismen (GAO) gegenereerd kan worden.

De bioreactor met de PAO cultuur die ook gebruikt is voor het onderzoek in hoofdstuk 5, werd langere voor langere termijn bedreven, om te beoordelen of de enzymen die nodig zijn voor de anoxische fosfor opname tot expressie konden komen onder verschillende operationele condities. De bioreactor werd bedreven onder anaërobe anoxische-oxische omstandigheden met variaties in (i) het synthetische medium (ii) de slibverblijftijd (SRT) (iii) de influent P / CZV verhouding, de nitraat dosering (iv) en (v) de aerobic SRT. Ondanks deze pogingen, werd er onder geen enkele van de operationele omstandigheden een aanzienlijke anoxische fosfor opname waargenomen. Bovendien werd er tijdens een activiteitstest waargenomen dat fosfor onder anoxische condities afgegeven werd, wat aangeeft dat PAO niet in staat zijn om PHA te oxideren voor de generatie van energie met behulp van nitraat.

Dit onderzoek geeft aan dat sulfaat reducerende bacteriën in de AWZI kunnen profileren aangezien de inhibitie reversibel is gedurende de recirculatie van het slib door anaerobe-anoxisch-oxische condities. De tot dusver bestaande kennis met betrekking tot het effect van sulfide op het anaerobe-oxische PAO metabolisme wordt door dit onderzoek vergroot. De onderzoeksresultaten suggereren dat de filamente bacteriën *Thiothrix caldifontis* een belangrijke rol kunnen vervullen in het biologisch verwijderen van fosfaat. Het werpt de vraag op of PAO in staat is energie te genereren uit nitraat respiratie en te gebruiken voor anoxische fosfaat opname. De door dit onderzoek verworven resultaten kunnen toegepast worden om de stabiliteit van het EBRP proces onder anaerobe-anoxische-oxische condities beter te begrijpen, met name gedurende blootstelling aan sulfide.

1

Introduction

1.1. Background

1.1.1. Importance of phosphorus removal from wastewater

The exponential increase in urbanization and industrialization of the 19th century led to an increase in the amount and complexity of wastewater generated by cities. During the 19th century, untreated wastewater was disposed of into surface water bodies. This practice increased the concentration of nitrogen and phosphorus in the environment, which increased the growth of bacteria and algae. The exponential growth of bacteria and algae (eutrophication) resulted in the utilization and subsequent depletion of oxygen (hypoxia), causing the death of aquatic life. As nitrogen can be fixed from the atmosphere by different organisms, phosphorus is the key nutrient to be removed in order to avoid eutrophication of water bodies (Yeoman et al., 1988).

In Europe up to 70 % of the phosphorus potentially discharged into the water bodies can be found in wastewater, whereas the other 30 % is generated by agriculture, aquaculture and industry activities (EEA, 2005). Thus, it is important to remove the phosphorus at the wastewater treatment plants in order to prevent eutrophication. Phosphorus can be removed from the wastewater either with the use of chemicals (e.g. iron dosage) or by biological means. The enhanced biological removal of phosphorus (EBPR) is a worldwide implemented process, whereas the phosphorus is removed by the sludge waste of microorganism capable to store phosphorus beyond its metabolic needs (Barnard, 1975). These polyphosphate accumulating organisms (PAOs) potentially can stored up to 0.38 mgP/mgVSS (Schuler et al., 2003) compared to 0.023 mgP/mgVSS which are used for growth by ordinary heterotrophic organisms (OHOs) (Metcalf & Eddy et al., 2003). PAOs normally proliferate in wastewater treatment plants (WWTP), which have anaerobic and oxic conditions, and recirculate the sludge to anaerobic conditions. Nevertheless, other microbes known as glycogen accumulating organism (GAOs) that do not contribute to the biological removal of phosphorus, can grow under similar conditions (Mino et al., 1987). GAOs are normally undesired on EBPR as

2

normally they are associated with the failure of the biological phosphorus removal process (Satoh et al., 1994).

1.1.2.   Saline wastewater

The world population will reach a maximum of 9.22 billion by 2075 (Chaime, 2004). Proportionally, the amount of food and water required to maintain the health and well-being of the population will increase. Due to population growth and depletion of potable water sources, the word "water stress" could become more common worldwide in the near future. Traditional solutions to alleviate the water scarcity like water saving, rainwater harvesting and fresh water transport from transboundary catchments will be insufficient to cope with the increasing fresh water demand (Karagiannis et al., 2008). Nevertheless, from the water available on the Earth, seawater comprises around 96% of the total. This characteristic makes seawater an apparently infinite source of water. Until now, seawater use has been limited to the industry for cooling systems, in fire departments for fire extinguishing, and only after desalination for direct human consumption (Lopez-Vazquez et al., 2009a). However, even if promising, the low efficiency of seawater desalination (reaching 50 % efficiency under the best case scenario), and relatively higher investment and operational costs (of up to €1.56/m³) makes this technology practically unsuitable to several countries (Karagiannis et al., 2008).

Alternatively, the use of saline water (e.g seawater, brackish water) as secondary quality water for sanitation can reduce up to 40 L/p.e. of fresh water demand (Lee et al., 1997). The use of seawater for sanitation has provided satisfactory results in terms of reduction of fresh water consumption, (ii) energy savings, and (iii) reduction of gas emissions in wastewater treatment systems in Hong Kong for more than 50 years (Li et al., 2005; Tang et al., 2007).

There are several concerns related to the collection, distribution and treatment of seawater such as: (i) pre-treatment of seawater, (ii) cross connections, (iii) corrosion of the distribution lines, (iv) deterioration of toilets, and (v) wastewater treatment in addition to the capital and maintenance cost generated (Tang et al., 2007). Nevertheless, most of these issues can be mitigated by: (i) electro chlorination, (ii) simple electro-conductivity measurements, (iii) better pre-treatment of seawater to reduce corrosion of distribution lines, (iv) cost-effective sulphate-based reduction processes for the saline wastewater generated (Ekama et al., 2011). Besides an increase in salinity, the use of seawater for sanitation purposes would increase the concentration of sulphate in the wastewater treatment plants.

Sulphate rich wastewater (containing up to 500 mg SO_4^{2-}/L) can be generated due to: (i) discharge of sulphate into the WWTP by industrial effluents (Sears et al., 2004), (ii) use of sulphate based chemicals in drinking water process (e.g. aluminium sulphate, Bratby, 2016), (iii) seawater and/or groundwater (rich in sulphate) intrusion (van den Brand et al., 2014a), and (iv) the use of seawater as secondary quality water (e.g. cooling, toilet flushing) (Lee et al., 1997). During sewage conveyance and in the anaerobic stages of a wastewater treatment (e.g. anaerobic sewerage and/or reactors) sulphate could be reduced to sulphide (H_2S/HS^-), which is inhibitory to the different microbial process of a WWTP (Comeau et al., 1986; Koster et al., 1986). Sulphide might cause microbial inhibition due to either direct inhibition of the unionized form of sulphide (dihydrogen sulphide, H_2S, which is able to pass through the cell membrane and reduce the intracellular pH)(Comeau et al., 1986; Koster et al., 1986), or precipitation of key micro-nutrients with sulphide (like copper, cobalt or iron) decreasing their bioavailability to cover the microbial metabolic requirements (Bejarano Ortiz et al., 2013; Zhou et al., 2014).

1.2. Scope of the thesis

Due to the different processes that can generate sulphate rich wastewaters (with up to 500 mg SO_4^{2-}/L), its occurrence in the inlet of wastewater treatment plants (WWTP) is likely to happen. Sulphate can be reduced to sulphide during sewage conveyance or in the different anaerobic tanks of the wastewater treatment plants (Bentzen et al., 1995; Poinapen et al., 2009). Certainly, these wastewater streams require proper treatment prior to discharge into surface water bodies. While COD and nitrogen removal can be satisfactorily achieved, the biological phosphorus removal process seems to be severely affected by the presence of sulphide (Comeau et al., 1986; Lau et al., 2006). The problem exacerbates if it is taken into account that phosphorus has been pointed out as the main nutrient that drives eutrophication in surface water bodies (Yeoman et al., 1988). Nevertheless, recently, Wu et al. (2014) was able to successfully couple the biological phosphorus removal with a sulphate reduction process. While promising, the operational conditions (e.g micro-aeration of 12 h and SRT above 68d) cannot be directly applicable to existing WWTP performing EBPR. Moreover, the effluent phosphorus concentrations were above the common discharge standards criteria for surface water bodies (of less than 1 mgPO$_4$-P/L). In later studies, Guo et al. (2016) solved the problem of an extended aeration, but the reactor still operates at high SRT (65 ± 12 d) and the removed phosphorus per amount of biomass (0.027 mg P/mgVSS) is close to the theoretical value used for growth (0.023 mg P/mg VSS; Metcalf & Eddy et al., 2003). Thus, it is still not clear until which concentration of sulphide is possible to maintain the biological removal of phosphorus.

This PhD studies aims to get a better understanding about the effects of sulphide on the dominant microbial populations involved in EBPR. This can contribute to improve and secure the satisfactory biological removal of phosphorus, when is exposed to the sulphide produced in the sewage (short term exposure) or at the WWTP (long term exposure).

1.3. Literature Review

1.3.1. Differences in polyphosphate accumulating organisms (PAOs)

Past research had identified and proposed different microorganism capable to store phosphorus beyond its growth requirements (known as poly-phosphate accumulating organisms; PAO). Early studies suggested that *Acinetobacter* as the main PAO responsible for the biological removal of phosphorus in wastewater treatment plants (Fuhs et al., 1975; Wentzel et al., 1986). Nevertheless, Bond et al. (1995) defined the community structure of a phosphorus removal sequencing batch reactor (SBR) and observed that *Acinetobacter* was poorly represented, whereas a specie from the *Rhodocyclus* genus was more abundant. In a similar manner, Hesselmann et al. (1999) applied 16s rRNA targeted molecular methods and observed a high abundance of *Rhodocyclus* in two phosphorus removing SBR. They suggested that this specie of *Rhodocyclus* is a new PAO, and they named it *Candidatus* Accumulibacter phosphatis (Hesselmann et al., 1999). *Candidatus* Accumulibacter phosphatis was further sub classified in two main clades (I and II) and several subclades, based on polyphosphate kinase gene (ppK1 and ppK2) (McMahon et al., 2002; Seviour et al., 2003; He et al., 2007; Peterson et al., 2008). Interestingly, the metagenomic analysis of the *Candidatus* Accumulibacter phosphatis subclades suggest several differences in their ability to generate energy from different electron acceptors (nitrate, nitrite, oxygen) (García Martín et al., 2006; Flowers et al., 2013; Skennerton et al., 2014).

During anaerobic conditions, *Candidatus* Accumulibacter phosphatis are capable to store volatile fatty acids (VFAs), such as acetate (HAc) or propionate (HPr), as poly-β-hydroxyalkanoates (PHA) (Comeau et al., 1986). The storage of VFA as PHA requires energy as ATP and a source of reducing equivalents (NADH) (Wentzel et al., 1986). *Candidatus* Accumulibacter phosphatis gain most of the energy needed from the hydrolysis of polyphosphate, which results in an increase in the concentration of soluble phosphate in the bulk liquid (Comeau et al., 1986; Wentzel et al., 1986). The source of reduction equivalents has been proposed to be provided either by glycogen consumption (glycolysis) (Mino et al ., 1987), the partial use of the TCA cycle (Comeau et al., 1986) or the combination of glycolysis and the TCA cycle (Pereira et al., 1996). In the presence of an electron acceptor (e.g. oxygen, nitrate and nitrite), *Candidatus* Accumulibacter phosphatis utilize the stored PHA as carbon and energy source to replenish their storage pools of polyphosphate and glycogen, to grow and for maintenance purposes (Comeau et al., 1986; Wentzel et al., 1986).

In later studies, Kong et al. (2005) observed that a community of *Actinobacteria* (close related to *Tetrasphera*) was capable to store phosphorus like *Candidatus* Accumulibacter phosphatis. Nevertheless, in contrast to the anaerobic metabolism of *Candidatus* Accumulibacter phosphatis, the *Actinobacter* present in the systems of Kong et al. (2005) were not able to store VFA as PHA. Later on, Nguyen et al. (2011) suggested that *Tetrasphera* (or *Actinobacter*) were capable to store acetate and ferment glucose, though it was not possible to identify the carbon storage compound of these organisms.

Recently, a new organism capable to store phosphorus beyond their growth requirements was detected in the cost of Namibia (*Thiomargarita namibiensis*). Schulz et al. (2005) suggested that *Thiomargarita namibiensis* was one of the organisms responsible to form phosphorite deposits in marine sediments. They observed that, during anaerobic conditions, *Thiomargarita namibiensis* used their internal storage pools of nitrate and phosphate to oxidize sulphide into sulphur and store poly-sulphur (Poly-S). While acetate triggered this anaerobic metabolism, it was not possible to observe any PHA inclusion and instead they suggested that acetate was stored as glycogen. During the presence of an electron acceptor, *Thiomargarita namibiensis* generate energy from Poly-S and glycogen to replenish their storage pools of poly-phosphate (Poly-P) (Schuler, 2005).

Brock et al. (2011) observed that a marine *Beggiatoa* strain was capable to store phosphate above its growth requirements. Contrary to *Candidatus* Accumulibacter phosphatis, the anaerobic phosphate release of the marine *Beggiatoa* culture was not affected by the addition of VFA. Thus, it was suggested that the anaerobic phosphate release was used to generate energy only for maintenance purposes, which increased accordingly to the concentration of sulphide. During oxic conditions, the stored poly-sulphur (Poly-S) was proposed to be used as a source of energy to replenish the phosphate storage of *Beggiatoa* (Brock et al., 2011).

Therefore, it seems that the ability to store phosphate during famine conditions and used it as energy source during feast conditions is widely distributed among different bacteria. However, due to their different metabolisms their proliferation and dominance in WWTP would depend on the operational conditions, configuration of the WWTP, and characteristics of the wastewater treated.

1.3.2.   Identification of PAO in WWTP

As previously discussed, there exist different organisms capable to store phosphorus above their growth requirements that potentially play a role on the biological removal of phosphorus at WWTP. In order to assess the relative abundance of each PAO in WWTP, analysis based on bio volume such as Fluorescence *in situ* hybridization (FISH), can be used. FISH analysis are performed with the help of rRNA-targeted probes for microbial identification, which emits an specific fluorescence signal under the microscope (Amman, 1995). Table 1 shows the potential PAOs active in wastewater treatment plant with the FISH probes that can be used for their identification.

Using FISH, Wagner et al. (1994) observed that the originally proposed PAO (*Acinetobacter*) was not abundant in full scale EBPR, which is in agreement with the lab-based observations of Bond et al. (1995). In a similar manner, to the best of our knowledge the involvement of *Thiomargarita namibiensis* and *Beggiatoa* in the biological removal of phosphorus at wastewater treatment plants has not been observed.

Both *Tetrasphera* and *Candidatus* Accumulibacter phosphatis have been observed to be active in WWTP (Saunders et al., 2003; Nguyen et al., 2011). *Tetrasphera* has been observed mainly in WWTP in Denmark (Nguyen et al., 2011), while the relative abundance of *Candidatus* Accumulibacter phosphatis has been correlated with good EBPR removal in WWTP with different configurations and located in 4 continents (Kong et al., 2002; Zilles et al., 2002; Saunders et al., 2003; He et al., 2005; Wong et al., 2005). As *Candidatus* Accumulibacter phosphatis seems to be more spread among WWTP, this study will focus mainly on *Candidatus* Accumulibacter phosphatis as one of the main PAO.

Table 1-1 FISH probes used to identify potential PAO active in WWTP.

Probes	Specify	Reference
ACA652	Acinetobacter	Wagner et al. (1994)
PAO462, PAO651, PAO846	Most Accumulibacter	Crocetti et al. (2000)
Acc-1-444	C. Accumulibacter clade I	Flowers et al. (2008)
Acc-2-444	C. Accumulibacter clade 2	Flowers et al. (2008)
Actino-221[a], Actino-658[a]	Actinobacteria	Kong et al. (2005)
Tet3-654, Tet2-842, Tet2-831, Elo1-1250, Tet1-266	Tetrasphera	Nguyen et al. (2011)
BEG811	Beggiatoa	Macalady et al. (2006)
Not found	Thiomargarita namibiensis	

a It requires a helper probe

1.3.3. Functional differences of *Candidatus* Accumulibacter phosphatis

Biological nutrient removal treatment plants aim to remove phosphorus and nitrogen from wastewater. As nitrate is an electron acceptor that can be used instead of oxygen for the generation of energy, past studies focused on the ability of *Candidatus* Accumulibacter phosphatis (hereafter referred as PAO) to use nitrate for the uptake of phosphorous (Kerrn-Jespersen et al., 1993; Kuba et al., 1993). Later studies, based on laboratory and pilot scale experiments, showed that anoxic phosphorus uptake was possible (Kuba et al., 1997a, 1997b; Kim et al., 2013). Nevertheless, they observed that the ability to use nitrate as electron acceptor depended on the PAO clade (Ahn et al., 2001a, 2002; Zeng et al., 2003b). García Martín et al. (2006) suggested that one of the PAOs, which could not use nitrate as electron acceptor was PAO clade IIA as it did not possess the nitrate reductase enzyme (*nar*). Nevertheless, PAO IIA contained the whole pathway necessary to denitrify from nitrite onwards (García Martín et al., 2006). Similar findings were reported by Skennerton et al. (2014), who did not observed the nitrate reductase enzyme (*nar*) in clades IIF, IIA, IA and IC but observed the nitrate periplasmic dissimilatory enzyme (*nap*). Based on metagenomic analysis and lab-observations it is believed that PAOI can denitrify whereas PAOII cannot, though some of the findings are contradictory (Carvalho et al., 2007; Lanham et al., 2011; Tayà et al., 2013; Ribera-Guardia et al., 2016; Saad et al., 2016).

Besides the difference in the anoxic metabolism of PAO clades, Welles et al. (2015) show that the poly-phosphate content in PAO affected differently the acetate uptake of PAO I and II. Even though only a limited number of studies differentiate among the PAO clades, it is clear that each clade of *Candidatus* Accumulibacter phosphatis performed slightly different metabolic activities.

This PhD thesis focuses on the effects of sulphide generated in sewer and WWTP in the biological removal of phosphorus. Thus, due to the potential ability of PAO I to denitrify and the importance of anoxic dephosphatation in WWTP, an enriched culture of PAO I was used during this study.

1.3.4. Factors affecting EBPR under saline conditions
1.3.4.1. Salinity

The different microbial processes at wastewater treatment plants would be affected by an increase in salinity. When the bacteria are exposed to a saline environment, they are affected by an increase in the osmotic stress and ionic strength (Roesser et al., 2001; Sleator et al., 2002). An increase in the ionic strength can denature the enzymes and proteins (Dale et al., 1983) and affect the lipid composition of the cell (Galinski et al., 1994). An increase in the osmotic stress causes that the water contained inside the bacteria flows from the cytoplasm to the solution until the bacteria dehydrates (Brown, 1990). In order to prevent dehydration, bacteria have developed two osmotic equilibrium mechanisms: i) the salt in cytoplasm mechanism and ii) the compatible solute mechanism (Galinski et al., 1994; Sleator et al., 2002). Recently Welles et al. (submitted) demonstrated that a good biological phosphorus removal can be achieved up to 3.5% salinity. Besides an increase in the osmotic stress and ionic strength, an increase in salinity can increase the concentration of sulphate in the WWTP.

1.3.4.2. Sulphate reduction

When treating saline wastewater, COD/SO_4^{2-} ratios around 0.6 and lower can be observed (Lau et al., 2006). In theory, the conversion of 1 g of sulphate would require 0.66 g COD (Liamleam et al., 2007). The reduction of sulphate can be done by autotrophic and/or heterotrophic sulphate reduction bacteria (SRB). Autotrophic SRB use CO_2 and H_2 as carbon and electron sources, respectively. On the other hand, heterotrophic SRB use organic compounds, like carbon, as electron sources (Liamleam et al., 2007).

Under anaerobic conditions, SRB will oxidize the organic matter and reduce SO_4^{2-}, producing hydrogen sulphide (Equation 1.1) (Tchobanoglous et al., 2003). Under such conditions, SRB will compete with anaerobic bacteria for the use of organic matter as carbon source. The outcome of this substrate competition will depend on the kinetics of bacterial growth, maximum specific growth rates, and substrate affinity (Liamleam et al., 2007). Under high sulphate concentration, and therefore low COD/SO_4^{2-} ratios, no methane gas production is usually observed indicating that SRB can outcompete methanogenic organisms under these conditions (Lu et al., 2011).

(Equation 1.1)

$$Organic\ matter + SO_4^{2-} \rightarrow S^{2-} + 2H_2O + CO_2 \rightarrow S^{2-} + 2H^+ \rightarrow H_2S$$

11

Sulphate reduction is commonly undesired in the treatment of wastewater due to: (i) possible corrosion (ii) interference with other biological/chemical process, and (iii) health and safety risk (Londry et al., 1999; Okabe et al., 2005). Nevertheless, sulphate reduction can also be beneficial when applied to wastewater treatment as sulphide can be used for: (i) heavy metal removal through precipitation (Lewis, 2010), (ii) autotrophic denitrification (Carmen et al., 2013), and (iii) reduction of pathogens (Abdeen et al., 2010). Furthermore, due to the slow growth and floc stability of the sulphate reducers is possible to achieve: (i) low biomass production (Lens et al., 2002), and (ii) granular sludge formation (van den Brand et al., 2014b).

Yamamoto et al. (1991) and Baetens et al. (2001) assessed the long-term effects of sulphate reduction on an EBPR system. In both cases, the system failed due to the increase of suspended solids in the effluent, which was caused due to the proliferation of filamentous bacteria. However, as pointed out by Daigger et al. (2015) this could be easily solved with the use of a membrane bioreactor. On the other hand, the effect of sulphide (and potential inhibition) on the metabolisms of PAO has not been fully reported.

During the competition for carbon source between SRB and PAO, it is expected that SRB consume more complex carbon sources (e.g. lactate, glucose; Zhao et al., 2008, 2010; Cao et al., 2012) whereas PAO consume the acetate and/or propionate generated from the fermentation process (Satoh et al., 1992; Oehmen et al., 2004). Moreover, the presence of PAO in the system could be favoured if the SRB community is mainly formed by incomplete oxidizing bacteria (e.g *Desulfovibrio, Desulfobulbus, Desulfomicrobium*)(Hao et al., 2014).

1.3.4.3. Sulphide

Sulphide is formed by the reduction of sulphate under anaerobic conditions. Sulphate reduction can occur during the anaerobic conveyance of wastewater in the sewage system or under the different anaerobic zones of the WWTP. Sulphide might cause microbial inhibition due to either direct inhibition of the unionized form of sulphide (dihydrogen sulphide, H_2S, which is able to pass through the cell membrane and reduce the intracellular pH)(Comeau et al., 1986; Koster et al., 1986), or precipitation of key micro-nutrients with sulphide (like copper, cobalt or iron) decreasing their bioavailability to cover the microbial metabolic requirements (Bejarano Ortiz et al., 2013; Zhou et al., 2014).

Sulphide affects differently the anaerobic and aerobic bacteriological processes, occurring in WWTP. It has been observed that 50 to 125 mgH_2S-S/L can inhibit 50% of the methane production (Chen et al., 2008). On the other hand, the effect of sulphide on aerobic processes seems to be more severe. Jin et al. (2013) reported that 32 mg H_2S-S/L caused 50% decrease in Anammox activity, whereas Bejarano Ortiz et al. (2013) observed that 2.6±0.3 mgH_2S-S/L and 1.2±0.2 mgH_2S-S/L caused 50% inhibition of the ammonia and nitrite oxidation activities in nitrifying cultures, respectively.

Only a few studies have focused on the effects of sulphide on the anaerobic metabolism of PAO. Comeau et al. (1986) observed that the addition of sulphide under anaerobic conditions led to an increased phosphate release, suggesting that phosphate was released to re-establish the intracellular pH after the disassociation of sulphide inside the cell. Similarly, Saad et al. (2013) reported that the anaerobic acetate uptake rate of PAO decreased around 50 % at 60 mg H_2S-S/L and observed 55 % higher anaerobic P-release, potentially associated to a detoxification process. However, no studies have assessed the effects of sulphide on the aerobic and anoxic metabolisms of PAO. Furthermore, it is not clear whether and to which extent the effects of sulphide are reversible.

1.4. Problem statement and objectives

Due to an increase in the water demand due to population growth and industrialization, it is expected that saline wastewater will be more common in the near future. One of the salts contain in saline wastewater is sulphate, which can reach concentrations up to 500 mg SO_4^{2-}/L. Under anaerobic conditions sulphate can be reduce into sulphide. Sulphide has shown to be inhibitory for the different microbial process of wastewater treatment plants. Therefore, it is necessary to study the effects of sulphide on the biological nutrient removal. Whereas some efforts had been done in assessing the effects of sulphide on the nitrogen and carbon removal, little is known about the effects of sulphide on EBPR. Therefore, this PhD thesis focuses to assess the effect of sulphide on EBPR systems. As the sulphide potentially generated in the sewage can vary greatly according to the rain and dry seasons, it is assumed that sulphide could cause a shock effect on EBPR (short-term effects). On the contrary, in case sulphate reduction occurs in the anaerobic tanks of WWTP, sulphide will continuously affect EBPR (long-term effects). However, the sulphate reduction activity on WWTP would depend in the effect of electron acceptors on sulphate reducing bacteria, as the sludge is commonly recirculate through anaerobic-anoxic-oxic conditions. Even more, past researchers had pointed out that the effect of sulphide on aerobic respiratory bacteria is more severe when compared with nitrite/nitrate respiratory bacteria. This, suggest that sulphide could affect differently the aerobic and anoxic metabolism of PAO. With the above mention in mind, the following specific objectives of this PhD research are:

- To assess the effects of electron acceptors on the activity of sulphate reducing bacteria.

- To evaluate the short-term exposure effects of sulphide on the anaerobic-oxic and anaerobic-anoxic metabolism of PAO

- To investigate the long-term effects of sulphide on the anaerobic- oxic and anaerobic-anoxic metabolism of PAO and potential microbiology selection or adaptation.

1.5. Research approach

Due to the potential ability of *Candidatus* Accumulibacter phosphatis clade I (PAO I) to denitrify and in order to compare the effects of sulphide on the anaerobic- anoxic - aerobic metabolism of PAO, an enriched culture of PAO I was used. In the first stage it was assessed the potential effect of different electron acceptors on the activity of sulphate reducing bacteria (SRB) (figure 1.1). To do so the activity of an enrich biomass of SRB was measured after its exposure during 2h to a similar concentration of oxygen, nitrate or nitrite to the ones observe in WWTP. In the second stage, two sequencing batch reactors (SBR) enriched with PAO I were operated either under anaerobic-aerobic or anaerobic-anoxic-aerobic conditions, to be subsequently exposed to sulphide. In order to differentiate between the sulphide formed in the sewage and WWTP, the enriched PAO I cultures were exposed to either short-term (hours) or long-term (weeks) to sulphide.

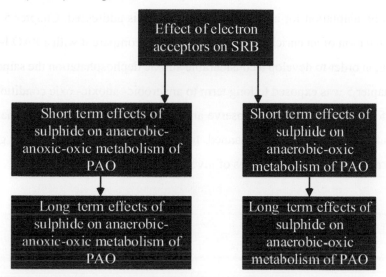

Figure 1.1.- Diagram of general approach of this research.

1.6. Outline

The present thesis comprises 7 chapters. The current chapter (chapter 1), contains a brief literature review relevant for the problem statement of this research. As pointed out on the research approach, first it was addressed the effect of electron acceptor on sulphate reducing bacteria. Such effect was assumed similar to the one potential caused due the recirculation of sludge through oxic and anoxic tanks. These findings can be found in chapter 2. On chapter 3 the effects caused by the suddenly exposure of sulphide (short-term), which could be form in the sewage, on the anaerobic-oxic metabolism of an enrich culture of PAO I were addressed. The study focused mainly in the effect of sulphide on the phosphate uptake and release rate as in the ammonia consumption rate of PAO I. Later on chapter 4, the effect of sulphide which is potential generated in the anaerobic tanks of the WWTP (long-term), was assessed on the anaerobic-oxic metabolism of an enrich culture of PAO I. The long term effects of sulphide were evaluated in like manner as in chapter 3, according to the phosphate uptake rate, phosphate release rate and the ammonia consumption. In chapter 4 the possibility of a microbial selection or adaptation for a PAO sulphide tolerant is addressed. Chapter 5 assess the anoxic dephosphatation of an enrich culture of PAO I, and compare it with a PAO I- GAO culture. In chapter 6, in order to develop a considerable anoxic dephosphatation the same PAO I culture used in chapter 5 was exposed to long term to anaerobic- anoxic- oxic conditions. As in neither chapter 5 or 6 were possible to observe any considerable anoxic dephosphatation, the effect of sulphide was not assessed as planned. In the last chapter (chapter 7) the general conclusions of this research as the future lines of investigations are discussed.

2

Effect of electron acceptors on sulphate reduction activity at WWTP

2.1. Highlights

- Oxygen is the most inhibiting electron acceptor to the sulphate reduction activity.

- Inhibition is more pronounced when sulphate reduction bacteria (SRB) are fed with acetate.

- The activity of sulphate reduction bacteria (SRB) may be prevented by applying an anaerobic contact time shorter than 0.4 h in conventional BNR systems.

Adapted from

Rubio-Rincon, F.J., Lopez-Vazquez, C.M., Welles, L., van den Brand, T.P.H., Abbas B, van Loosdrecht , M., Brdjanovic, D. (submitted) Effect of electron acceptors on sulphate reduction activity at WWTP.

2.2. Abstract

The concentration of sulphate in wastewater can vary from 10 to 500 mg SO_4^{2-}/L. Under anaerobic conditions, sulphate could potentially be reduced to sulphide by sulphate reducing bacteria (SRB). The generation of sulphide is generally undesired. Past research indicated that SRB are inhibited by the presence of electron acceptors (O_2, NO_3 and NO_2). However, the contact time and concentration used in those studies are by far higher than the ones that may be found in sewage treatment plants. Thus, this research aims to understand how the different electron acceptors commonly present in biological nutrient removal (BNR) systems can affect the proliferation of SRB. For this purpose, a culture of SRB was enriched in a sequencing batch reactor. Once enriched, the SRB were exposed for 2 h to similar concentrations of electron acceptors like those observed in BNR systems. Their activity was assessed using three different types of electron donors (acetate, propionate and lactate). Oxygen was the most inhibiting electron acceptor regardless the carbon source used. After exposure to oxygen and when feeding acetate, a residual inhibition of the SRB activity was observed for 1.75 h. Thereby, only 60% of the original sulphate reduction activity was recovered. The proliferation of SRB may be prevented in conventional BNR if the anaerobic contact time is shorter than 0.4 h. These results can be used to implement strategies to control the growth of sulphate reducers.

Keywords: Sulphate reduction activity, sulphate reducing bacteria, wastewater treatment, electron acceptor inhibition.

2.3. Introduction

Sulphate-rich wastewater (containing up to 500 mg SO_4^{2-}/L) can be generated due to: (i) discharge of sulphate into the WWTP by industrial effluents (Sears et al., 2004), (ii) seawater and/or groundwater (rich in sulphate) intrusion (van den Brand et al., 2014c), (iii) use of sulphate chemicals in drinking water production (e.g. aluminium sulphate, Bratby, 2016) and (iv) the use of seawater as secondary quality water (e.g. cooling, toilet flushing) (Lee et al., 1997).

Heterotrophic dissimilatory sulphate reduction can occur in sulphate-rich waters under anaerobic conditions at COD/SO_4 ratios higher than 0.66 mg COD/ mg SO_4^2(Liamleam et al., 2007; Muyzer et al., 2008). Under anaerobic conditions, sulphate reducers can compete with anaerobic bacteria for a wide range of carbon donors such as: glucose, lactate, propionate, acetate, butyrate, ethanol, and even bicarbonate, among others (Muyzer et al., 2008). The degradation of carbon by sulphate reducers can be divided in two groups: (i) complete degradation into carbon dioxide, and (ii) partial degradation to acetate (Liamleam et al., 2007).

Hydrogen sulphide, the end product of sulphate reduction, is commonly undesired in the treatment of wastewater due to: (i) possible corrosion, (ii) interference with other biological/chemical process, and (iii) health and safety risks to workers (Londry et al., 1999; Okabe et al., 2005). Nevertheless, sulphate reduction can also be beneficial when applied to wastewater treatment as sulphide can be used for: (i) heavy metal removal through precipitation (Lewis, 2010), (ii) autotrophic denitrification (Kleerebezem et al., 2002) accompanied with reduced BOD requirements for N-removal, and (iii) reduction of pathogens (Abdeen et al., 2010).

To repress the formation of sulphide, past studies have focused on the development of measures that can inhibit the sulphate reduction process (or SRB). One way to inhibit SRB activity is by avoiding the creation of anaerobic conditions through the addition of oxygen or nitrate (Dilling, Waltraud et al., 1990; Bentzen et al., 1995). Another approach is through the addition of metabolic inhibitors such as molybdate and nitrite (Nemati et al., 2001).

Oxygen has shown to be toxic for many anaerobic bacteria, such as sulphate reducing bacteria (Lens et al., 2001). Still, sulphate reducers can to endure the (short-term or partial) exposure to oxic conditions by: (i) the oxygen respiration at the expense of poly-glucose (Kjeldsen et al., 2005; Dolla et al., 2006), (ii) adherence to biofilms where the gradients reduce their exposure to oxygen or other electron acceptors, (Lens et al., 2001), and (iii) their potential symbiosis with oxygen oxidizing organisms (e.g. sulphide oxidizing bacteria) (van de Ende et al., 1997; Xu et al., 2012, 2014). Moreover, once the conditions become anaerobic again, sulphate reducers can recover their activity (Kjeldsen et al., 2005; Nielsen et al., 2008).

Likewise oxygen, nitrate and/or nitrite have been applied to suppress the sulphate reduction process and/or oxidize the sulphide generated back to elemental sulphur or sulphate (Bentzen et al., 1995; Mohanakrishnan et al., 2008, 2009). During the long-term exposure to nitrate, García De Lomas et al. (2006) observed the growth of sulphide denitrificans (*Thiomicrospira*) in an enriched sulphate reducing biomass. Thus, García De Lomas et al. (2006) suggested that the lower sulphide production observed during the presence of nitrate was not due to inhibition of the sulphate reduction process but it was related to the sulphide consumed for denitrification using nitrate.

During such autotrophic denitrification process, others researchers observed an accumulation of nitrite in the media (Hubert et al., 2005; Barton et al., 2007). Barton et al. (2007) suggested that the inhibition due to nitrate of sulphate reducing bacteria observed in the past could be related to nitrite formation. Later studies showed that nitrite was able to suppress the reduction of sulphite (SO_3^{2-}) to sulphide (HS^-) (Barton et al., 2007; Mohanakrishnan et al., 2008). Regardless the addition of nitrate or nitrite, once these compounds were consumed, after certain time the sulphate reduction process ressummed reaching a reduction activity similar to the pre-inhibition levels (Okabe et al., 2005; Mohanakrishnan et al., 2008).

Since the addition of oxygen, nitrate and/or nitrite is not irreversibly inhibitory to sulphate reducction bacteria, their activity is expected to resumme once the conditions become again anaerobic. Despite that there are several studies on SRB inhibition caused by their exposure to electron acceptors (oxygen, nitrate, nitrite), there is still a need to understand the potential sulphate reduction activity under similar conditions to those transient conditions observed in wastewater treatment plants (WWTPs). Therefore, this research aims to understand how different anaerobic/anoxic and oxic contact times affect the inhibition and recovery of SRB and hypothesizes which conditions in a WWTP (anaerobic, anoxic, and oxic) can be manipulated to either promote or inhibit the growth and activity of sulphate reducers in WWTP.

2.4. Material and methods
Reactor operation

A culture of sulphate reduction bacteria (SRB) was enriched in a double jacketed Applikon reactor (Delft, The Netherlands) with a working volume of 2.5 L. Activated sludge (500 mL) from WWTP Nieuwe Waterweg (Hoek van Holland, The Netherlands) was used as inoculum. The bio reactor was operated in cycles of 6h with an effective 5h anaerobic reaction time, 30 min settling and 30 min effluent removal. In order to ensure anaerobic conditions (assumed to occur at redox levels lower than -400mV), nitrogen gas was sparged during the first 20 min of operation and a double water lock was installed and connected to the headspace. One bottle was filled with NaOH to capture the sulphide produced and the other one with $NaSO_3+CoCl_2$ to remove the potential oxygen from intrusion. During the effluent withdrawal phase, half of the working volume was removed to reach a hydraulic retention time (HRT) of 12h. The solids retention time (SRT) was controlled at 15d by removing 41 mL of mixed liquor sludge at the end of the anaerobic phase. The pH was adjusted at 7.6 ± 0.1 through the addition of 0.4M HCl and 0.4M NaOH. Temperature was controlled at $20\pm1°C$ with a water bath. The redox level was monitored continuously online and it fluctuated between -400 and -480 mV. Sulphate (SO_4-S), sulphide (H_2S-S), total suspended solids (TSS) and volatile suspended solids (VSS) were measured twice per week. When no significant changes in these parameters were observed for at least 3 SRT (45d), it was assumed that the system was in pseudo steady-state conditions.

Media

The media was separated into two bottles of 10L (COD and mineral sources). Each bottle (containing of the media solutions) was sterilized at 110 °C for 1h. The mixed media fed into the reactor contained per litre: 93 mg NaOAc•$3H_2O$ (43 mg COD), 29 μL of propionic acid (44 mg COD), 216 μL of lactic acid (237 mg COD), 107 mg NH_4Cl (28 mg NH_4-N), 112 mg $NaH_2PO_4•H_2O$ (25 mg PO_4-P), 1.24 gr $MgSO_4•7H_2O$ (498 mg SO_4^{2-}), 14 mg $CaCl_2•2H_2O$ (4 mg Ca^+), 36 mg KCl (19 mg K^+), 1 mg yeast extract, 2 mg N-allylthiourea (ATU) and 300 μL of trace element solution prepared according to Smolders et al. (1994).

Control batch activity test

Batch activity tests were performed in 500 mL double-jacketed reactors with a working volume of 400 mL. 200 mL of biomass from the parent reactor (±900 mg VSS/L) was used to conduct each control batch test. After the sludge transfer from the parent to the batch reactor, the waste of sludge of the parent reactor was adjusted to compensate for the withdrawal of biomass. Nitrogen gas was sparged at the bottom at 10 L/h during 30 min prior to the start and during the conduction of the control batch activity test to ensure the creation of anaerobic conditions. Three carbon sources (acetate, propionate, and lactate) were added separately and tested. The batch tests were performed for 6 h similar to the operation of the parent reactor. The pH and temperature were controlled at 7.6±0.1 and 20±1°C. The sludge was constantly magnetically stirred at 300 rpm. TSS and VSS were measured at the start and end of the test. Acetate, propionate, lactate, sulphide and sulphate were analysed along the execution of all the experiments.

Batch activity tests executed under the presence of electron acceptors

The residual effects of three electron acceptors (2.7 mg O_2/L, 15 mg NO_3-N/L, and 10 mg NO_2-N/L) on the sulphate reduction process using three different electron donors (acetate, propionate and lactate) were assessed separately. In each test, 200mL of biomass from the parent reactor (±900 mg VSS/L) were transferred to double-jacketed reactors with a working volume of 400 mL. 200 mL of mineral solution (free of organics) was added in combination with one of the electron acceptors. Each test lasted for two hours. Thereafter, the corresponding electron acceptor was removed by washing the sludge and a reversibility test was conducted. The reversibility tests were performed following the same conditions like the control batch tests.

Addition and removal of electron acceptors in the inhibitory batch tests

Oxygen was constantly measured (±2.7 mg O_2/L) and controlled by a mixture of compressed air (10 L/h) and nitrogen gas (20L/h; van de Ende et al., 1997) throughout the corresponding inhibitory batch tests. Once the inhibitory tests concluded (after 2h), nitrogen gas was sparged during 20 min at 10 L/h until oxygen was no longer detected.

Nitrate and nitrite were added for a concentrated stock solution (containing 1 g NO_x-N/L) to reach a concentration of 15 mg NO_3-N/L and 10 mg NO_2-N/L, respectively, in the beginning of the inhibitory tests. In order to prevent the occurrence of denitrification, nitrate and nitrite were measured at the start and end of the tests. At the end of the tests, the previously added nitrate or nitrite were removed. For this purpose, three washing steps were performed. Each washing step consisted of a settling phase of 20 min, removal of supernatant (±90% of the volume), and addition of a mineral solution similar to the one used in the control batch tests. After the three washing steps, this resulted in a final nitrate/nitrite concentration below detection levels (>0.1 mg NO_x-N/L). Nitrogen gas was sparged at the headspace continuously to avoid oxygen intrusion.

Analyses

Samples were filtered through 0.45 μm pore size filters (PDVF). In order to avoid the oxidation of sulphide, the samples used for sulphate and sulphide determination were kept in a 0.5 M NaOH solution (the reading values were corrected according to the dilution caused by the addition of the NaOH solution). All samples were measured in the subsequent 2 h after the tests concluded. Sulphate was measured by ion chromatography (IC) using a Dionex Ionpack AS4A-SC column (Dreieich, Germany). Nitrite, sulphide, TSS, and VSS were analysed as described in APHA (2005). Nitrate was measured according to ISO 7890/1 (1986). Acetate and propionate were measured using a Varian 430-GC Gas Chromatograph (GC) equipped with a split injector (200°C), a WCOT Fused Silica column (105°C) and coupled to a FID detector (300°C). Helium gas was used as carrier gas and 50μl of butyric acid as internal standard. Lactate was measured in a high-performance liquid chromatography (HPLC) using a Trace 2000 chromatograph (Thermo Electron S.P.A., Milan, Italy).

Determination of stoichiometric and kinetic parameters of interest

The net carbon consumption per sulphate reduction (mg COD/mg SO_4^{2-}) was calculated based on the total carbon consumption and sulphate reduction observed after a 1 hour of activity. All kinetic rates were calculated by linear regression as described in Smolders et al. (1995). The rates of interest were:

q_{Ac}:   Acetate consumption rate, in mg COD-Ac/gVSS.h

q_{Pr}:   Propionate consumption rate, in mg COD-Pr/gVSS.h

q_{Lac}:   Lactate consumption rate, in mg COD-Lac/gVSS.h

q_{COD}:   Organic carbon consumption rate, in mg COD /gVSS.h

$q_{SO4,Ac}$:   Sulphate reduction associated with acetate consumption, in mg SO_4-S /gVSS.h

$q_{SO4,Pr}$:   Sulphate reduction associated with propionate consumption, in mg SO_4-S /gVSS.h

$q_{SO4,Lac}$:   Sulphate reduction associated with lactate consumption, in mg SO_4-S /gVSS.h

Microbial characterization

Fluorescence in situ Hybridization (FISH) analyses were performed according to Amman (1995) to identify the presence of the microbial communities of interest. In order to target all bacteria, equal amounts of EUB 338, EUB338 II and EUB 338 III probes were mixed (EUB MIX) and applied (Halkjær et al., 2009). Most Desulfovibrionales and other Bacteria were targeted with the SRB385 probe and most Desulfobulbus with the DBB660 probe (Baumgartner et al., 2006; Muyzer et al., 2008). Vectashield with DAPI was used to amplify the fluorescence, avoid the fading and staining all organisms (Halkjær et al., 2009). Biomass quantification was performed through image analysis of 20 random pictures taken with an Olympus BX5i microscope and analysed with the software Cell Dimensions 1.5. The standard error of the mean was calculated as described by Oehmen et al. (2010). Denaturing gradient gel electrophoresis (DGGE) was performed as described by Bassin et al. (2011). After analysis the reads library was imported into CLC genomics workbench v7.5.1 (CLC Bio, Aarhus, DK) and (quality, limit=0.05) trimmed to an minimum of 200bp and average of 284bp. A build-it SILVA 123.1 SSURef Nr99 taxonomic database was used for BLASTn analysis on the reads under default conditions. The top result was imported into an excel spreadsheet and used to determine taxonomic affiliation and species abundance.

2.5. Results

Performance of the parent reactor

In the parent reactor, all carbon sources were consumed under anaerobic conditions. Lactate was consumed at a rate of 185 mg COD-Lac/(gVSS.h), while propionate and acetate were consumed at the slower rates of 24.7 mg COD-Pr/(gVSS.h) and 20.7 mg COD-Ac/(gVSS.h), respectively (Figure 2.1). Sulphate was reduced at a rate of 26 mg SO$_4$-S/(gVSS.h) during the first two hours of reaction. After this time, it was not possible to observe any considerable change in the sulphate/sulphide concentration in the liquid phase. The COD-conversion/SO$_4$-reduction conversion ratio observed in the first 2 hours was 0.64 mg COD/mg SO$_4{}^{2-}$.

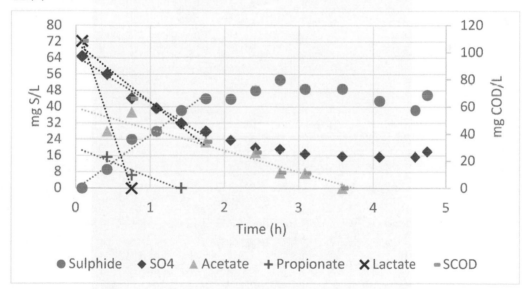

Figure 2.1.- Sulphide (circle), sulphate (diamond), acetate (triangle), propionate (cross), lactate (X) and soluble COD (dash) profiles of the SRB enrichment culture observed during a typical cycle in the parent reactor.

According to the FISH image analyses, the microbial community targeted with the EUB mix probe covered about 79 ± 6 % EUB/DAPI (Figure 2.2B) of the whole cells that reacted with DAPI (Figure 2.2A). The bacterial community targeted with the EUB mix probe consisted of 88±4% SRB385/EUB sulphate reducing bacteria (SRB) (Figure 2.2C), from which 96 ± 9% DBB660/SRB385 belonged to the genera *Desulfobulbus* (Figure 2.2D). These FISH analyses are in line with the results gathered by Denaturing Gradient Gel Electrophoresis (DGGE), which shows a clear presence of *Desulfobulbus* and *Desulfobacter* (Figure 2.3).

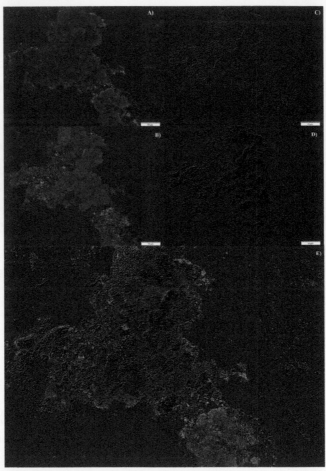

Figure 2.2.- FISH microbial characterization of the biomass present in the parent reactor: A) DAPI (all living organism), B) EUB MIX (all bacteria), C) SRB385 (most sulfate reducers) and D) DBB660 (Desulfobulbus). Figure E shows an overlap of Figures A to D.

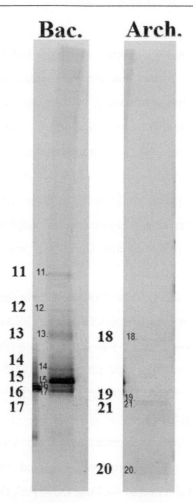

Nr. Band	Access number	Order	Family	Genus
11	HQ688408.1	*Flavobacteriales*	*Flavobacteriaceae*	N.C
12	DQ443910.1	*Clostridiales*	*Lachnospiraceae*	*Clostridium*
15	NR_028830.1	*Desulfobacterales*	*Desulfobacteraceae*	*Desulfobacter*
16	AJ012591.1	*Desulfobacterales*	*Desulfobulbaceae*	*Desulfobulbus*
17	AY297798.1	*Desulfuromonadales*	*Geobacteraceae*	*Geobacter*
18	KJ436577.1	*Clostridiales*	*Lachnospiraceae*	*Clostridium*
20	JX225727.1	*Desulfobacterales*	*Desulfobacteraceae*	N.C.
21	JQ177807.1	*Desulfuromonadales*	*Geobacteraceae*	*Geobacter*

N.C.- not classified

Figure 2.3.- DGGE banding pattern and phylogenetic tree of the biomass enriched in the parent reactor.

Control batch tests

The control batch tests fed with lactate showed a faster COD consumption than the batch test fed with propionate or acetate. Lactate was consumed at a rate of 282.6 mg COD-Lac/gVSS.h. During the consumption of lactate, the formation of propionate and acetate was observed. Thus, the overall organic soluble COD consumption was 172 mg COD/(gVSS.h) (Figure 2.4A). Once lactate was completely consumed, propionate and acetate were consumed at rates of 49.6 mg COD-Pr/(gVSS.h) and 35.6 mg COD-Ac/(gVSS.h), respectively. In line with the COD consumption, it was observed that the sulphate reduction associated with lactate consumption occurred at a faster rate compared to the sulphate reduction associated with acetate consumption (Table 2.1). The overall COD/SO_4^{2-} consumption ratio was 1.79 mg COD/mg SO_4^{2-}.

In the control batch test fed with propionate, acetate formation was observed. While propionate was consumed at a rate of 84.6 mg COD-Pr/(gVSS.h), acetate was formed at a rate of 16 mg COD-Ac/(gVSS.h) and sulphate was reduced at a rate of 24 mg SO_4-S/(gVSS.h) (Figure 2.4B). During the experimental time propionate was not completely consumed and acetate consumption was not observed. The overall COD/SO_4^{2-} ratio in the propionate control batch test was 0.97 mg COD/mg SO_4^{2-}.

The control batch test fed with acetate showed the slowest carbon consumption and sulphate reduction observed among the three tests executed with the three different carbon sources. Acetate was consumed at a rate of 40.4 mg COD-Ac/(gVSS.h) together with 19.3 mg SO_4-S/(gVSS.h) reduction. The COD/SO_4 consumption ratio observed was the smaller one among the three carbon sources used (of 0.70 mg COD/mg SO_4^{2-}).

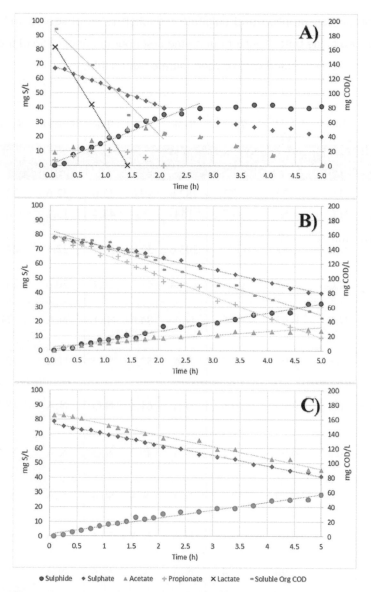

● Sulphide ◆ Sulphate ▲ Acetate ＋ Propionate ✕ Lactate ▬ Soluble Org COD

Figure 2.4. Profiles observed in the control tests showing the concentrations of sulphide (circle), sulphate (diamond), acetate (triangle), propionate (cross), lactate (X), soluble organic COD (dash) profiles in the control test performed with A) lactate, B) propionate or C) acetate as carbon source.

Table 2.1.- Carbon consumption and sulphate reduction rates observed with the different carbon sources fed to the batch reactor during the first hour of conduction of the control tests and the inhibiting tests executed with oxygen, nitrate, and nitrite.

		Lag phase	COD/SO$_4$	q_{COD}	q_{Lac}	q_{Pr}	q_{Ac}	$q_{SO4,Lac}$	$q_{SO4,Pr}$	$q_{SO4,Ac}$
	Units	hours	mgCOD/ mg SO$_4$	mgCOD/ gVSS.h	mgCOD/gVSS.h			mgSO$_4$-S/gVSS.h		
Lactate as carbon source	Control	N.O.	1.79	172	282.6	49.6	35.6	32.6	32.2	14.4
	Oxygen test	N.O.	1.58	83.7	154	33	48.4	15.1	18.1	11.2
	Nitrate test	N.O.	1.57	113.2	203.6	32.2	23.7	22.1	24.7	12.4
	Nitrite test	N.O.	1.42	94.4	199.4	41.7	19.4	20.6	25.9	10.9
Propionate as carbon source	Control	N.O.	0.97	68.6	N.A	84.6	N.O.	N.A	24	N.O
	Oxygen test	0.6	1.07	41.5	N.A	52.8	N.O	N.A	12.9	N.O
	Nitrate test	0.4	0.90	49.5	N.A	61.3	N.O	N.A	18.3	N.O
	Nitrite test	0.4	0.97	37.1	N.A	49.2	N.O	N.A	12.7	N.O
Acetate as carbon source	Control	N.O.	0.70	40.4	N.A	N.A	40.4	N.A	N.A	19.3
	Oxygen test	1.75	0.83	26.1	N.A	N.A	26.1	N.A	N.A	10.4
	Nitrate test	0.6	0.90	19.4	N.A	N.A	19.4	N.A	N.A	7.2
	Nitrite test	0.4	0.83	33.6	N.A	N.A	33.6	N.A	N.A	13.4

N.O. – Not observed ; N.A..- Not applicable

Oxygen inhibition test

Like in the control batch tests, the experiment fed with lactate showed a faster organic COD uptake and sulphate reduction than the experiments fed with propionate or acetate (Table 2.1). Compared to the control batch test, the exposure to oxygen decreased the lactate consumption and the sulphate reduction rates to 154 mg COD-Lac/(gVSS.h) and 15 mg SO$_4$-S/(gVSS.h), respectively. Similarly, the propionate formed was consumed at a slower rate of 33 mg COD-Pr/(gVSS.h) coupled to a sulphate reduction rate of 18 mg SO$_4$-S/(gVSS.h). However, the concentration of acetate formed due to the incomplete lactate oxidation was higher in the oxygen inhibition test than in the control test (61 and 51 mg COD-Ac/L, respectively). Thereafter, the acetate consumption rate was higher than in the control test (48 and 35 mg COD-Ac/(gVSS.h), respectively). Interestingly, the sulphate reduction associated with the consumption of the acetate formed was slightly slower than in the control test (of 11 mg SO$_4$-S/(gVSS.h) and 14 mg SO$_4$-S/(gVSS.h), respectively).

When propionate was fed as carbon source the exposure to oxygen caused a longer lag phase in the propionate consumption of 0.6 hours. After the exposure of the biomass to oxygen, it was observed that both the propionate consumption and the associated sulphate reduction rates were 48% and 46% slower than the rates observed in the corresponding control test (Table 2.1). Similar to the control test fed with propionate as carbon source, propionate was not completely consumed and the acetate formed was not consumed either (Annex 2.1E).

In the tests conducted with the addition of acetate as carbon source, it was not possible to observe any acetate consumption for 1.75 h after the exposure of the biomass to oxygen. Once the activity of the biomass resumed, the acetate consumption rate was 26.1 mg COD-Ac/(gVSS.h) and the sulphate reduction rate reached 10.4 mg SO$_4$-S/(gVSS.h).

Nitrate inhibition test

During the nitrate inhibition batch test, the residual activity of SRB after the exposure to nitrate was assessed. It was possible to observe that the residual activity of SRB after the exposure to nitrate was higher when lactate was fed (Table 2.1).

The consumption of lactate and sulphate started immediately after the conditions where switched from anoxic to anaerobic. Lactate and sulphate were consumed at rates equivalent to 38% and 32% of the rates observed in the control test (Table 2.1). A similar residual effect was observed in the consumption of the propionate and acetate formed in the test (65 % and 66%, respectively). On the other hand, the sulphate reduction associated with the propionate and acetate consumption was higher but still lower than the rate observed in the control (76 % and 86%, respectively).

When propionate was used as carbon source it was not possible to observe any consumption of propionate during the first 0.4 h of the reaction time. After 0.4 h, the propionate consumption rate was 61.3 mg COD-Pr/(gVSS.h) coupled with 18.3 mg SO_4-S/(gVSS.h) reduction. In a similar way like in the oxygen inhibition and control tests, propionate was not completely consumed during the reaction time. Thus, it was not possible to observe any consumption of the acetate formed due to the incomplete oxidation of propionate.

A more severe effect was observed on the nitrate inhibitory test fed with acetate. In this experiment, it was not possible to observe any consumption of acetate or sulphate reduction in the first 0.6h of reaction. After this period, acetate was consumed at rate of 19.4 mg COD-Ac/(gVSS.h) together with 7.2 mg SO_4-S/(gVSS.h) reduction.

Nitrite inhibition test

During the nitrite inhibition test fed with lactate, lactate was consumed at 199.4 mg COD-Lac/(gVSS.h) while the propionate and acetate formed were consumed at 41.7 mg COD-Pr/(gVSS.h) and 19.4 mg COD-Ac/(gVSS.h), respectively (Table 2.1). Like in the previous batch tests, the sulphate reduction related to the propionate consumption was higher than the sulphate reduction related to lactate consumption (Table 2.1). On the other hand, the sulphate reduction coupled with the consumption of acetate was around half of that observed with either lactate or propionate (10.9 mg SO_4-S/(gVSS.h)).

In the nitrite tests fed with propionate, it was not possible to observe any activity in the first 0.4 h. Afterwards, a slower propionate consumption of 49.2 mg COD-Pr/(gVSS.h) coupled with 12.7 mg SO_4-S/(gVSS.h) was observed. A similar lag-phase was observed when acetate was used as carbon source (0.4h). Nevertheless, the recovery of the acetate consumption rate and the coupled sulphate reduction rate were higher (33.6 mg COD-Ac/(gVSS.h) and 13.4 mg SO_4-S/(gVSS.h, respectively).

2.6. Discussion

Characterization of biomass in the bioreactor

Base on the net COD and SO_4^{2-} transformations inside the reactor (108 mg COD/L and 45 mg S/L), the COD/SO_4^{2-} conversion ratio was around 0.80 mg COD/mg SO_4^2. This ratio is higher than the one of 0.64 mg COD/mg SO_4^{2-} calculated based on the first two hours of reaction time and is closer to the theoretical COD consumption per sulphate reduced (Choi et al., 1991).

The theoretical ratio of 0.66 mg COD/ mg SO_4^{2-} and the quantification of the SRB using FISH analysis can be used to estimate the observed biomass yield using Equation 2.1 (Henze et al., 2008). In this case, the observed biomass yield of SRB would be between 0.072 mg VSS/mg COD (0.10 mg VSS/mg SO_4^{2-}) and 0.086 mg VSS/mg COD (0.13 mg VSS/mg SO_4^{2-}) considering that either all or only 84% of the organic COD fed was consumed by SRB. These values are in line with the ones reported for other SRB cultures enriched in similar conditions than the ones applied in this study (pH 7.6, 20°C, using acetate and propionate as carbon source) (Lens et al., 2002; van den Brand et al., 2014c). Therefore, based on the microbial characterization and conversion ratios observed in the parent reactor, sulphate reducing bacteria seems to have been the dominant organisms present in the parent reactor (Figures 2.1, 2.2 and 2.3).

$$Y_{obs} = \frac{MxVss}{Qi*Sbi*SRT} = \frac{VSS*V}{Qi*Sbi*SRT}$$ Equation 2.1

Where:

VSS:	volatile suspended solids	mg VSS
Qi:	influent flow	L/d
Sbi:	Soluble biodegradable COD	mgCOD/L
SRT:	Solids retention time	d

This research aims to understand the residual activity of SRB after exposure to different electron acceptors. Therefore, as previous research had pointed out that the effects of toxic and inhibitory compounds on SRB are dependent on their carbon source consumed (Maillacheruvu et al., 1996) it is important to assess the consumption of each carbon source with regard to the sulphate reduction activity. As observed in Figure 2.1, part of the lactate fed (27%) was fermented to propionate and acetate. According to Oyekola et al. (2012), the half saturation constant of lactate oxidizing organisms is 0.12 gLac/L, whereas for lactate fermenters is 3.30 gLac/L. Thus, the lower concentration of lactate added in this study (0.23 g COD/L) should have been beneficial for the oxidation process over the fermentation process of lactate, which was not the case. Taken into account the remained acetate formed (19 mg COD_{Ac}/L), sulphide production, and Equation 2.6 is possible to estimate that 49% of the total lactate was incompletely oxidized with sulphate to acetate. This is in line with the results presented by Dar et al. (2008) who suggested that incomplete oxidizers SRB overcompete complete oxidizers. According to the Gibbs free energy presented in table 2.2, sulphate reducing bacteria can generate as twice as much energy during the incomplete oxidation of lactate compared with its complete oxidation (-160.3 and -84.9 Kj/mol S, respectively). This might explain why in this set of experiments most of the lactate was incompletely oxidize into acetate by SRB. While the remained lactate (23%) might be oxidized into carbon dioxide.

Residual inhibitory effect of different electron acceptors on SRB

Table 2.1 shows that none of the electron acceptors tested caused an irreversible inhibition of the SRB activity. This is line with past research which reported that SRB can fully recover their activity observed before inhibition once the conditions become anaerobic again (van de Ende et al., 1997; Greene et al., 2003; Kjeldsen et al., 2004; Okabe et al., 2005; Mohanakrishnan et al., 2008). Moreover, according to previous studies certain SRB poses the capability to respire on the different electron acceptors (e.g. detoxification or metabolic process) used in this study (van de Ende et al., 1997; Greene et al., 2003; Hubert et al., 2005).

Table 2.2- Possible sulfate reduction reactions in an enrich SRB bioreactor.

Equation	$\Delta G_o^{'a}$ (Kj/reaction)	$\Delta G_o^{'}$ (Kj/mol S)
(2.2) $C_2H_3O_2^- + SO_4^{2-} \rightarrow HS^- + 2HCO_3^-$	-47,3	-47,3
(2.3) $4C_3H_5O_2^- + 3SO_4^{2-} \rightarrow 3HS^- + 4HCO_3^- + 4C_2H_3O_2^- + H^+$	-151,3	-50,42
(2.4) $4C_3H_5O_2^- + 7SO_4^{2-} \rightarrow 7HS^- + 12HCO_3^- + H^+$	-340,5	-48,64
(2.5) $3C_3H_5O_3 \rightarrow C_2H_3O_2 + 2C_3H_5O_2 + CO_2 + H_2O$	-170,0	N.A.
(2.6) $2C_3H_5O_3^- + SO_4^{2-} \rightarrow HS^- + 2HCO_3^- + 2C_2H_3O_2^- + H^+$	-160,3	-160,3
(2.7) $2C_3H_5O_3^- + 3SO_4^{2-} \rightarrow 3HS^- + 6HCO_3^- + H^+$	-254,9	-84,9

a .- $\Delta G_o^{'}$ values taken from Thauer et al., (1977)

Effects of aerobic exposure time on SRB

After the biomass was exposed to oxygen, the residual sulphate activity was similar independently of the carbon source used (Table 2.1). This is in agreement with the observations of Cypionka (1994) who concluded that some species of sulphate reducers (e.g. *Desulfovibrio, Desulfobulbus*) were capable to survive a continuous exposure to oxygen. As observed in Figures 2.2 and 2.3, at least one of these species was present in the bioreactor. On the other hand, in the past it was observed that the presence of sulphide induced the formation of hydrogen peroxide and superoxide radicals, increasing the toxicity of oxygen to anaerobes

(Cypionka, 1994). However, in this case nitrogen gas was sparged at the bottom of the rector before the biomass was exposed to oxygen, hence, sulphide should have been stripped out as opposed to the findings of Cypionka et al. (1985). Thus, it is assumed that the residual effect observed on SRB in this research is solely due to the oxygen exposure and no to the combined effect of sulphide with oxygen.

Interestingly, the recovery of the organic carbon uptake rate was lower with lactate (48%) than with propionate (60%) or acetate (65%), whereas the sulphate reduction associated with the carbon consumption was not considerably different (53±2%). According with Kjeldsen et al. (2004), the reduction of the sulphate reduction activity observed after the exposure to oxygen could be partially caused by a reduction of substrate produced by anaerobic bacteria. However, as in this case SRB were capable to directly use lactate, this mechanism can be discarded. Therefore, despite that other anaerobic bacteria that may consume lactate were more severely affected than SRB by the exposure to oxygen; it seems that also the exposure to oxygen partially inhibited the sulphate reducing activity.

Despite that, similar inhibition activities (of 53%) were observed in the oxygen tests fed with either lactate, propionate or acetate, the inactivation time (Lag phase) was different in each test. The longest lag phase observed was 1.75 h (when feeding acetate). In contrast, the use of lactate as carbon source did not present any lag phase. This suggests that the exposure to oxygen should promote the growth of SRB able to use lactate over those that use acetate as carbon source. This in line with the studies of Lens et al. (1995) who identified that SRB capable to use lactate as carbon source were present in aerobic wastewater treatment plants.

Following a similar approach as the one used in the cycle of the parent reactor and using the equations displayed in Table 2.2, it is possible to calculate the use of carbon source related to sulphate consumption. This approach shows that the relative fermentation of lactate increased from 19% to up to 35% once the biomass was exposed to oxygen. On the other hand, the relative percentage of lactate oxidation related to sulphate consumption remained similar in both cases (37%), whereas the lactate oxidized to carbon dioxide decreased (42% to 25%

on the control batch and oxygen stress test, respectively). Thus, as SRB which oxidized acetate or propionate were more inhibited due to oxygen and the fraction of lactate fermented increased, the exposure of biomass to oxygen could result in the accumulation of VFA in the anaerobic selectors. Therefore, the accumulation of VFA could result beneficial for other microbial process such as denitrification or the biological removal of phosphorus. Nevertheless, the sulphide produced by SRB could hinder the anaerobic and more severely the aerobic metabolism of PAO (Rubio-Rincon et al., chapter 3).

Effects of anoxic exposure time on SRB

Past research had suggested the use of nitrate or nitrite as inhibitory compounds for sulphate reduction activity (Bentzen et al., 1995; Greene et al., 2003; García De Lomas et al., 2006). It is believed that nitrite and not nitrate is the compound that actually inhibits the dissimilatory sulphate reduction from sulphite onwards (Hubert et al., 2005; Okabe et al., 2005; Barton et al., 2007). However, in this research sulphide was observed to be formed as a product of sulphate reduction, in biomass previously exposed to nitrate or nitrite possibly because these compounds were not present during the sulphate reduction process. This might suggest that the effect of nitrate or nitrite in the reduction of sulphate is reversible, as previously reported (Greene et al., 2003; Kjeldsen et al., 2004; Mohanakrishnan et al., 2008).

In the past experiments, nitrate showed to affect more severely the recovery of the sulphate reduction activity on acetate-consuming SRB than in lactate/propionate-SRB (36% and 76%, respectively). This is in agreement with Maillacheruvu et al. (1993), who observed a higher tolerance to toxicants of SRB able to oxidize lactate or glucose compared to SRB that oxidize acetate or propionate. In the same way, the soluble organic carbon consumption rate in the nitrate inhibitory tests fed with lactate or propionate recovered up to 72% of their pre-inhibition activity, while the one fed with acetate recovered solely 48% of its initial activity (Table 2.1).

Moreover, the period where it was not possible to observe neither carbon consumption nor sulphate reduction was 0.6 h when the system was fed with acetate. But, it resumed immediately when the system was fed with lactate, suggesting that fermentative and lactate SRB were active. Such lag phase suggest that acetate cannot be immediately consumed. Kjeldsen et al. (2004) suggested that the lag phase or period of inactivation was related to the different microbial communities. In their experiments, they suggest that the lag phase was caused by the inhibition of fermentative bacteria. Thus, the differences observed in these experiments could be due to the presence of different SRB with different capacities to tolerate the presence of electron acceptor. Greene et al. (2003) suggested that the nitrate reductase enzyme (Nrf) was widely distributed among SRB and could be used for detoxification processes. In that case, the ability of different SRB to express this enzyme (or the use of different carbon sources) might result in the different periods of inactivation observed in this research.

Due to the slower inactivation time when the biomass was exposed to either nitrate or nitrite compared to oxygen, likely SRB more tolerant to anoxic conditions can proliferate in biological nutrient removal systems without significantly affecting them as VFA can be available for denitrification and the biological removal of phosphorus. Furthermore, the long-term exposure to nitrate/nitrite may induce the growth of autotrophic denitrifiers (Reyes-Avila et al., 2004; García De Lomas et al., 2006; Wang et al., 2009a) that could grow coexisting with SRB hindering the inhibition caused by the addition of nitrate or nitrite.

2.7. Conclusion

The inhibition of the sulphate reduction process was more severe on sulphate reducing bacteria (SRB) able to use acetate, which suggest that SRB capable to use lactate as carbon source are more likely to proliferate in biological nutrient removal wastewater treatment plants (WWTP) due to their typical alternating anaerobic-anoxic-aerobic stages. While the biomass exposed to oxygen experienced a longer inactivation period, 53% of their sulphate reducing activity was recovered after 1.75 h. In order to minimize SRB grow and favour the VFA accumulation on anaerobic selectors, it is suggested to decrease the anaerobic contact time below 0.4 h.

3

Sulphide effects on the physiology of Candidatus Accumulibacter phosphatis Type I

3.1. Highlights

- Sulphide affects both the anaerobic and aerobic metabolism of PAO I.

- Sulphide inhibition on the metabolism of PAO I is partially reversible.

- Aerobic growth in PAOI was fully inhibited at sulphide concentrations >8 mgH_2S-S/L.

- At sulphide concentrations higher than 36mg H_2S-S/L, aerobic P-release was observed.

- The anaerobic processes were inhibited by 50 % at 22 mg H_2S-S/L

Adapted from
Rubio-Rincón F.J., Lopez-Vazquez C.M., Welles L., van Loosdrecht M.C.M., Brdjanovic D. (2016) Sulphide effects on the physiology of Candidatus Accumulibacter phosphatis Type I, Applied microbiology, and biotechnology.

3.2. Abstract

Sulphate rich wastewaters can be generated due to: (i) use of saline water as secondary quality water for sanitation in urban environments (e.g. toilet flushing), (ii) discharge of industrial effluents, (iii) sea and brackish water infiltration into the sewage, and (iv) use of chemicals, which contain sulphate, in drinking water production. In the presence of an electron donor and absence of oxygen or nitrate, sulphate can be reduced to sulphide. Sulphide can inhibit microbial processes in biological wastewater treatment systems. The objective of the present study was to assess the effects of sulphide concentration on the anaerobic and aerobic physiology of polyphosphate accumulating organisms (PAOs). For this purpose, a PAO culture, dominated by Candidatus Accumulibacter *phosphatis clade I* (PAO I), was enriched in a sequencing batch reactor (SBR) fed with acetate and propionate. To assess the direct inhibition effects and their reversibility, a series of batch activity tests were conducted during and after the exposure of a PAO I culture to different sulphide concentrations. Sulphide affected each physiological process of PAO I in a different manner. At 189 mgTS-S/L, VFA uptake was 55% slower and the phosphate release due to anaerobic maintenance increased from 8 to 18 mgPO$_4$-P/gVSS.h. Up to 8 mgH$_2$S-S/L, the decrease in aerobic phosphorus uptake rate was reversible (Ic$_{60}$). At higher concentrations of sulphide, potassium (>16 mgH$_2$S-S/L) and phosphate (>36 mgH$_2$S-S/L) were released under aerobic conditions. Ammonia uptake, an indicator of microbial growth, was not observed at any sulphide concentration. This study provides new insights into the potential failure of EBPR sewage plants receiving sulphate or sulphide rich wastewaters when sulphide concentrations exceed 8 mgH$_2$S-S/L, as PAO I could be potentially inhibited.

Keywords: Sulphide inhibition, Enhanced Biological Phosphorus Removal (EBPR), Poly-phosphate Accumulating Organisms (PAO), *Candidatus Accumulibacter phosphatis* Clade I

3.3. Introduction

To reduce eutrophication in surface water bodies, phosphate needs to be removed from wastewater by biological or chemical means in wastewater treatment plants (Yeoman et al., 1988; Henze et al., 2008). The biological removal of phosphate is carried out by microorganisms broadly known as polyphosphate accumulating organisms (PAOs) capable of storing phosphate beyond their biomass growth requirements as intracellular polyphosphate (poly-P) (Comeau et al., 1986; Mino et al., 1998). During anaerobic conditions, PAOs store volatile fatty acids (VFA) (e.g. propionate, acetate) as poly-hydroxy-β-alkanoates (PHA), which require additional energy and reduction equivalents. PAOs obtain most of the required energy (ATP) from the hydrolysis of intracellular poly-P, which results in the release of phosphate and cations (e.g. calcium, magnesium, potassium) (Comeau et al., 1986). Conversion of glycogen to PHA provides extra energy (ATP) and reducing equivalents (NADH) needed (Mino et al., 1998).

Thereafter, under aerobic or anoxic conditions (depending on the presence of oxygen, nitrate, and/or nitrite), PAOs consume the stored PHA to replenish their poly-P and glycogen storage pools, to grow, and for maintenance purposes (Comeau et al., 1986; Wentzel et al., 1986). Thus, in order to support the development of PAO and achieve enhanced biological phosphorus removal (EBPR) in a wastewater treatment plant (WWTP), mixed liquor activated sludge should be cycled through alternating anaerobic and aerobic/anoxic conditions and the influent should be directed to the anaerobic stage (Henze et al., 2008).

Sulphate rich wastewaters (containing up to 500mg SO_4^{2-}/L) can be generated due to: (i) discharge of sulphate into the WWTP by industrial effluents (Sears et al., 2004), (ii) use of sulphate based chemicals in drinking water production (e.g. aluminium sulphate)(Bratby et al., 2016), (iii) seawater and/or groundwater (rich in sulphate) intrusion (van den Brand et al., 2014a), and (iv) use of seawater as secondary quality water (e.g. cooling, toilet flushing) (Lee et al., 1997). During sewage conveyance and in the anaerobic stages of a wastewater treatment (e.g. anaerobic sewerage and/or reactors), sulphate could be reduced to sulphide (H_2S/HS^-) and inhibit different organisms (Comeau et al., 1986; Koster et al., 1986). Sulphide might cause microbial inhibition due to either direct inhibition of the unionized form of sulphide (dihydrogen sulphide, H_2S, which is able to pass through the cell membrane and reduce the intracellular pH) (Comeau et al., 1986; Koster et al., 1986) or precipitation of key micro-nutrients with sulphide (like copper, cobalt or iron) decreasing their bioavailability to cover the microbial metabolic requirements (Bejarano Ortiz et al., 2013; Zhou et al., 2014).

The sulphide inhibition effects on certain microorganisms have been already assessed. Chen et al. (2008), working with an anaerobic suspended sludge bioreactor, reported 50 % methanogenic inhibition at sulphide concentrations between 50 and $125mgH_2S$-S/L, whereas Koster et al. (1986) observed 50 % inhibition at $250mgH_2S$-S/L in an anaerobic granular sludge fed with acetate. Jin et al. (2013) reported that 32mg H_2S-S/L caused a 50 % decrease in Anammox activity, meanwhile Bejarano Ortiz et al. (2013) observed that 2.6 ± 0.3 mgH_2S-S/L and 1.2 ± 0.2 mgH_2S-S/L caused 50% inhibition of the ammonia and nitrite oxidation activities in nitrifying cultures, respectively. These observations suggest that the aerobic or anoxic metabolic activities appear to be more sensitive to the presence of sulphide than the anaerobic one.

So far only a few studies have focused on the effects of sulphide on the anaerobic metabolism of PAO. Comeau et al. (1986) observed that the addition of sulphide under anaerobic conditions led to an increased phosphate release, suggesting that phosphate was released to re-establish the intracellular pH after the disassociation of sulphide inside the cell. Similarly, Saad et al. (2013) reported that the anaerobic acetate uptake rate of PAO decreased around 50 % at 60mg H_2S-S/L and observed 55 % higher anaerobic P-release, potentially associated to a detoxification process. However, no studies have reported the effects of sulphide on the aerobic metabolism. Furthermore, it is not clear whether and to what extent the effects of sulphide on the metabolism of PAO are reversible.

Thus, the main objective of the present research was to study the short-term effects of sulphide on the anaerobic/aerobic physiology of an enriched PAO culture (dominated by Candidatus Accumulibacter *phosphatis* Clade *I*, hereafter PAO I). To assess the direct inhibition effects, batch tests were performed at different sulphide concentrations ranging from 48 to 189 mgTS-S/L (H_2S+HS^-) added at the start of the anaerobic stage. Once sulphide was depleted, in order to assess the reversibility of the inhibition effects, additional batch tests were performed with the same sludge immediately after the direct exposure tests. The findings will improve our understanding regarding the inhibiting effects of sulphide on PAOs and serve to develop strategies to overcome their deleterious effects on EBPR systems.

3.4. Materials and methods
Reactor operation

The biomass was enriched in a 3.0 L double-jacket Applikon reactor with a working volume of 2.5L (Delft, Netherlands). 500 mL of activated sludge from WWTP Nieuwe Waterweg (Hoek van Holland, The Netherlands) was used as inoculum. The reactor was operated in cycles of 6 h (2h 15 min anaerobic phase, 2h 15 min aerobic phase, 1h settling time and 30 min effluent removal). At the start of each cycle, 1.25 L of synthetic media was fed to the reactor (5min feeding), resulting in a hydraulic retention time (HRT) of 12h. Through the wastage of 78 mL of mixed liquor at the end of each aerobic phase, the solids retention time (SRT) was controlled at 8 d. The pH was kept at 7.6 ± 0.1 through the addition of 0.1M HCl and 0.4M NaOH. The dissolved oxygen (DO) concentration was maintained at around 20% of the saturation concentration through the automatic supply of compressed air or nitrogen gas. Temperature was externally controlled at 20±1°C. The ortho-phosphate, VFA, mixed liquor suspended solids (MLSS), and mixed liquor volatile suspended solids (MLVSS) concentrations were measured twice per week at the start and end of each phase (anaerobic/aerobic). When no significant changes in these parameters were observed for at least 3 SRT, it was assumed that the system was under pseudo steady-state conditions.

Synthetic media

The media was concentrated 10 times and separated in two bottles (carbon source and mineral solution). After dilution, the synthetic media fed to the reactor contained per litre 637 mgNaAc•$3H_2O$ (295mgCOD/L), 66.7 μL propionic acid (100 mgCOD/L), 107 mgNH$_4$Cl, 111 mgNaH$_2$PO$_4$•H_2O (25mgPO$_4$-P/L), 90 mgMgSO$_4$•$7H_2O$, 14 mgCaCl$_2$•$2H_2O$, 36 mgKCl, 1mg yeast extract, 20 mgN-allylthiourea (ATU) and 300μL of trace element solution prepared according to Smolders et al. (1994).

Batch activity tests

Batch tests were performed by duplicate in two jacketed reactors, each one with a working volume of 400 mL. When the biomass performance in the parent SBR was under steady-state conditions, 200 mL of sludge were transferred from the parent SBR to each batch reactor. Prior to the execution of each batch test, the media was sparged with nitrogen gas and adjusted to pH 7.6. The length of each batch test was composed of 75 min anaerobic and 125 min aerobic conditions. To ensure anaerobic conditions, nitrogen gas was sparged at the bottom of the batch reactors during feeding and thereafter to their headspace during the rest of the anaerobic phase. In the aerobic stage, compressed air was sparged from the bottom. Both gases were controlled at 10 L/h. The pH was kept at 7.6 ± 0.1 through the automatic addition of HCl and NaOH. As described elsewhere (Lopez-Vazquez et al., 2008), the oxygen consumption rates (OUR) were determined in 2-3 min time intervals at DO concentrations higher than 2 mg/L in a separate 10 mL unit equipped with an OXi 340i DO probe (WTW,Germany).

Direct inhibition H$_2$S batch tests

Direct inhibition batch tests were performed at different initial total sulphide (H$_2$S+HS) concentrations added at the start of the anaerobic phase. Sulphide concentrations ranged between 48 and 189 mgTS-S/L. Anaerobic synthetic media was diluted two times to keep a similar organic load to biomass ratio (F/M) like in the parent reactor and the pH adjusted to 7.6. Before the addition of media to the batch reactor, either 2.5, 5.0, 7.5 or 10 mL of concentrated sulphide (containing 3.2 g H_2S+HS^-/L) was added and the final volume adjusted to 200 mL. The pH of the media was adjusted to 7.6 prior to addition.

Reversible inhibition H₂S batch tests

Immediately after the completion of each direct inhibition test (once sulphide was no longer detected), the reversibility tests were conducted in the same reactor and with the same biomass previously exposed to a defined sulphide concentration. Thus, following the same approach like in the direct inhibition tests but excluding the addition of sulphide, anaerobic conditions were created through nitrogen gas addition, the synthetic media was added, and thereafter the sequential anaerobic-aerobic stages were conducted.

Analyses

Samples were filtered through 0.45 μm pore size filters. 50 μL of butyric acid as internal standard was added to the samples of acetate (HAc) and propionate (HPr) and stored in 1 ml sampling bottles. Iron (Fe^{2+}, Fe^{+3}), Potassium (K^+), Magnesium (Mg^{2+}), and Calcium (Ca^+) were stored in 0.5 % nitric acid solution. Orthophosphate (PO_4^{3-}-P), ammonia (NH_4^+-N), acetate (HAc) and propionate (HPr) were analysed within 2 hours after each batch test. Total sulphide (H_2S+HS^-) was measured immediately after sampling. Orthophosphate (PO_4^{3-}-P), total sulphide (H_2S+HS^-), total suspended solids (TSS), and volatile suspended solids (VSS) were analysed as described in APHA et al. (2005). Ammonia (NH_4^+-N) was measured according to NEN 6472 (1983). Acetate (HAc) and Propionate (HPr) were measured using a Varian 430-GC Gas Chromatograph (GC) equipped with a split injector (200 °C), a WCOT Fused Silica column (105°C) and coupled to a FID detector (300 °C). Helium gas was used as carrier gas. Iron (Fe^{2+}, Fe^{+3}), Potassium (K^+), Magnesium (Mg^{2+}), and Calcium (Ca^+) where measured in an Inductively Coupled Plasma, Mass Spectroscopy) (Thermo Scientific in Bremen, Germany).

Fluorescence *in situ* Hybridization (FISH)

To identify the dominant microbial communities in the sludge, Fluorescence *in situ* Hybridization (FISH) analyses were performed according to Amman (1995). To target all bacteria, the EUB MIX probe (mixture of EUB 338, EUB338 II and EUB 338 III probes) was applied (Amman, 1995). PAO were targeted with the PAOMIX probe (composed of probes PAO 462, PAO 651 and PAO846) (Crocetti et al., 2000). The presence of PAO clade I and clade II was estimated through the addition of probes Acc-1-444 (1A) and Acc-2-444 (2A, 2C, 2D) (Flowers et al., 2009). *Candidatus Competibacter phosphatis* was targeted with the GB probe (Kong et al., 2002). Defluvicoccus cluster 1 and 2 were identified with the TFO-DF215, TFO-DF618, DF988, and DF1020 probes (Wong et al., 2004;Meyer et al., 2006). Vectashield containing a DAPI concentration was used to amplify the fluorescence signal and stain all living organisms (Nielsen et al., 2009).

An estimation of the relative biomass fractions of the organisms of interest was performed by analysing 25 random FISH image fields taken with an Olympus BX5i microscope and analysed with the software Olympus Cell Dimensions 1.5 (Hamburg, Germany). The relative abundance of the organisms was estimated by expressing the relative surface area, that stained positive with the specific probes, with regard to the total surface area stained positive with DAPI (Flowers et al., 2009). The standard error of the mean was calculated as described elsewhere.

Stoichiometric and kinetic parameters of interest

The net P-released per VFA consumption ratio (P-mmol/C-mmol) was calculated based on the total P released observed at the end of the anaerobic phase and the total VFA consumed. The ratios mmol-K^+/P-mmol, mmol-Mg^{2+}/P-mmol, and mmol-Ca^+/P-mmol were calculated based on their net differences at the start and end of the anaerobic phase (anaerobic ratios), and at the start and end of the aerobic phase (aerobic ratios). The poly-P content in the biomass was not measured, but estimated based on the ash content, as described by Welles et al. (2015).

All kinetic rates were calculated by linear regression and expressed as mg compound per g of volatile suspended solids (VSS) per hour (mg/(gVSS.h)), as described in Smolders et al. (1995). The anaerobic rates of interest were:

q_{VFA}^{MAX}:   Maximum specific VFA consumption rate.

$q_{PO_4,AN}^{MAX}$:   Maximum specific total phosphate release rate.

$q_{PO_4,VFA}$:   Maximum specific phosphate release rate for VFA uptake, estimated as

$$q_{PO_4,VFA} = q_{PO_4,AN}^{MAX} - m_{PO_4,AN} \qquad \text{Equation 3.1.}$$

$m_{PO_4,AN}$:   Specific anaerobic phosphate release rate due to maintenance purposes (after VFA consumption).

m_{I,PO_4}: Specific anaerobic phosphate release rate for maintenance or detoxification purposes (after VFA consumption) associated to the presence of sulphide, estimated as the increase in phosphate release rate between the $m_{PO_4,AN}$ observed in the control test (at 0 mgH$_2$S-S/L) and in the different (direct and reversibility inhibition) tests exposed to sulphide (48 and 89 mgTS-S/L).

Similarly, the aerobic rates of interest were:

$q_{PO_4,Ox}^{Ini}$:   Maximum specific initial phosphate uptake rate, determined based on the phosphate uptake rate observed during the presence of sulphide (in the direct inhibition tests) or within the first 60 min of the aerobic phase (in the reversibility tests).

$q_{PO_4,Ox}^{Res}$:   Residual phosphate uptake rate, estimated based on the phosphate uptake rates measured once sulphide was not detected in the aerobic phase.

$q_{NH_x,Ox}$:   Specific ammonia uptake rate.

$m_{O_2,Ox}$:   Oxygen uptake rate measured at the end of the aerobic phase (associated to maintenance purposes) once phosphate uptake stopped and sulphide was not detected.

Inhibitory sulphide expressions

Mathematical expressions were developed to describe the direct inhibition effects of total sulphide on the anaerobic and aerobic phosphate profiles of PAO. Sulphide was modelled as total sulphide ($HS^- + H_2S$). The results from the direct inhibition tests, showing the effect of sulphide on the physiology of PAO I, were used to calibrate the model. Based on the phosphate profiles, three mathematical expressions were proposed and added to the model developed by Lopez-Vazquez et al. (2009). The anaerobic phosphate release rate due to detoxification caused by sulphide (m_{I,PO_4}) was described with the following expression (Equation 3.2):

$$m_{I,PO4}^{s} = m_{I,PO4}^{max} \cdot \frac{S_{H_xS}}{S_{H_xS} + K_{H_xS,AN}} \cdot X_{PAO} \qquad \text{Equation 3.2}$$

where:

$m_{I,PO4}^{s}$ Corresponds to the increase in the anaerobic P-release rate (presumably associated to the detoxification process) caused by sulphide, in mg PO_4-P/L.

$m_{I,PO4}^{max}$ is the maximum increase in the anaerobic P-release rate caused by sulphide, in mg PO_4-P/gVSS.h.

$K_{H_xS,AN}$ is the half saturation inhibition constant affecting the increased P-release rate associated to the presence of sulphide, in mg S/L.

S_{H_xS} is the sulphide concentration in the liquid phase, in mgS/L.

X_{PAO} is the fraction of PAO biomass, in gVSS.

For the description of the aerobic physiology, two different phases were considered: (i) the first one under the presence of sulphide, hereafter identified as directly inhibited activity, and (ii) a second one once sulphide was no longer detected, henceforward referred to residual activity. To describe the directly inhibited activity, the aerobic P-uptake rate was affected by the inclusion of an inhibition expression (Equation 3.3). For the description of the residual activity, which takes into consideration that the inhibition may be partially reversible, an empirical inhibiting expression for the aerobic P-uptake rate was proposed (Equation 3.4).

$$\frac{K_{I,Ox}}{K_{I,Ox}+S_{H_xS}}$$ Equation 3.3

$$\frac{K_{I,Ox}\cdot e^{S_{H_xS}^{ref}-S_{H_xS}^{Ini}}}{K_{I,Ox}\cdot e^{S_{H_xS}^{ref}-S_{H_xS}^{Ini}}+S_{H_xS}^{Ini}}$$ Equation 3.4

where,

$K_{I,Ox}$ is the half saturation inhibition constant of sulphide on the aerobic metabolism of PAO, in mg S/L.

$S_{H_xS}^{ref}$ is a reference sulphide concentration at which an irreversible inhibition of the aerobic phosphate uptake rate can occur, in mg S/L.

$S_{H_xS}^{Ini}$ is the initial sulphide concentration at the start of the aerobic phase, in mg S/L.

Aquasim was used to estimate the kinetic parameters used in the proposed expressions and to simulate the phosphate concentration profiles in the different experiments (Reichert, 1998). The percent error was calculated through the normalized root mean square deviation (NRSMD), as described by Oehmen et al. (2010).

3.5. Results

Sludge performance and microbial community in parent reactor

Prior to the execution of the batch inhibition tests, the parent reactor was operated for more than 200 days showing a stable pseudo steady-state performance with a VSS/TSS ratio of 0.59 gVSS/gTSS and complete phosphate removal at the end of the aerobic phase (Figure 3.1, i). All VFA was consumed within 10 minutes of the anaerobic phase at 534 mgCOD/gVSS.h, releasing phosphate at a rate of 377 mg PO_4-P/gVSS.h and reaching a P to VFA ratio of 0.76 P-mmol/C-mmol. In the presence of oxygen, phosphate was taken up at an initial rate of 57.9 mgPO$_4$-P/gVSS.h, consuming 0.54 mmol-O_2 per P-mmol taken up. FISH analyses were performed on day 217, showing that the biomass was composed of 97 ± 4 %

Candidatus Accumulibacter phosphatis (with regard to DAPI stained biomass), from which around 99 % belonged to Candidatus Accumulibacter phosphatis clade I (Figure 3.1, ii).

A control test was performed under the same conditions as the direct inhibition test (without the addition of sulphide). In the control test, all VFA (measured as COD) were consumed within 15 min, at a rate of 510 mgCOD/gVSS.h. During the anaerobic stage, 76 mg PO_4-P/L were released at a rate of 399 mgPO$_4$-P/gVSS.h ($q_{PO_4,AN}^{MAX}$). The stoichiometric P-to-VFA ratio was 0.78 P-mmol/C-mmol. In the aerobic stage, phosphate was taken up with a maximum specific P-uptake rate 68 mgPO$_4$-P/gVSS.h ($q_{PO_4,Ox}^{Ini}$), consuming 0.55 O$_2$-mmol per P-mmol taken up (Figure 3.2). The anaerobic/aerobic stoichiometry and kinetics of the control test were in the same range to those observed in the cycle of the parent reactor.

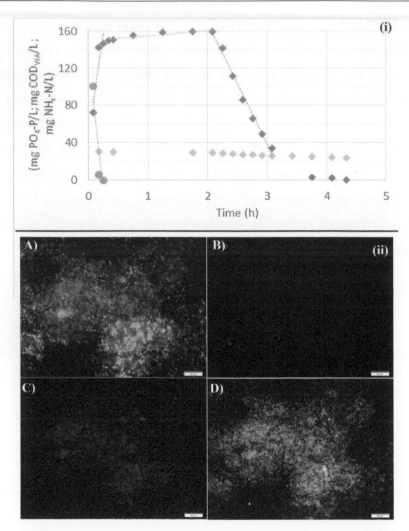

Figure 3.1.- Sludge characterization in the parent SBR displaying (i) the present profiles of carbon (blue circle), phosphate (orange triangle) and ammonia (as NH4-N) (yellow diamond) observed in a typical operating cycle under steady-state conditions, and (ii) microbial identification by Fluorescence in situ Hybridization (FISH) showing the microbial community in the sludge (bar indicates 20 µm): A) all living organism (DAPI) in green, (B) GAO (Cy5) in blue; C) PAO (Cy3) in red, and D) PAO I (Fluos) in yellow.

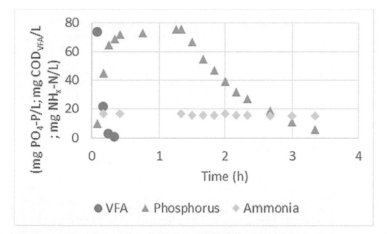

Figure 3.2.- Batch activity control test performed with PAO biomass enriched in the parent SBR showing the profiles of: VFA (measured as COD)(blue circle), phosphate (PO$_4$-P)(orange triangle), and ammonia (as NH$_4$-N)(yellow diamond).

Direct sulphide inhibition effects on the anaerobic and aerobic metabolism of PAO I

Under anaerobic conditions, the sulphide concentrations added in the beginning of each test were rather stable. Figure 3.3 shows the phosphate, VFA (as COD), ammonia, and sulphide profiles during and after a short-term exposure to the different concentrations of sulphide assessed that increased from 0 to 42 mg TS-S/L (8 mgH$_2$S-S/L), 86 mgTS-S/L (16 mgH$_2$S-S/L), 115 mgTS-S/L (22 mgH$_2$S-S/L), and 189 mg TS-S/L (36 mgH$_2$S-S/L). In the different anaerobic stages studied, the VFA uptake rates decreased progressively (Figure 3.4) from 510 to 227 mgCOD/gVSS.h as the sulphide concentration increased. Nevertheless, in all cases VFA was taken up within the length of the anaerobic phases. The maximum phosphate release rate ($q_{PO_4,AN}^{MAX}$) showed a similar trend decreasing from 399 mgPO$_4$-P/gVSS.h to 119 mgPO$_4$-P/gVSS.h. On the contrary, the anaerobic maintenance P-release rates ($m_{PO_4,AN}$) increased from 8 to 18 mgPO$_4$-P/gVSS.h h as the sulphide concentration reached 189 mgTS-S/L (36 mgH$_2$S-S/L).

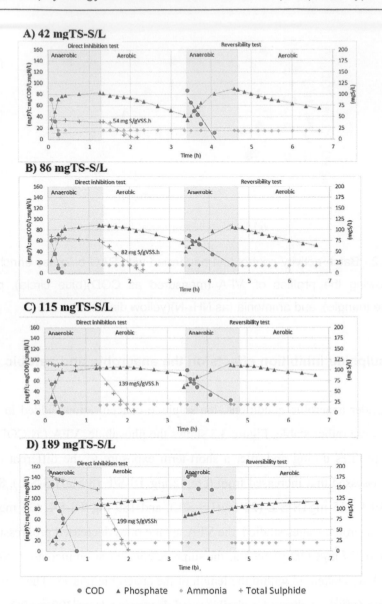

Figure 3.3.- Phosphate (PO4-P), VFA (as COD), ammonia (as NH4-N) and total sulphide (H2S+HS- as S) profiles during the direct inhibition and subsequent reversibility tests executed at: A) 42 mgTS-S/L (8 mgH2S-S/L), B) 86 mgTS-S/L (16 mgH2S-S/L) , C) 115 mgTS-S/L (22 mgH2S-S/L), and D) 189 mgTS-S/L (36 mgH2S-S/L).

Regardless the sulphide concentration, in the first 60 min of the aeration phase sulphide dropped below detection limits (Figure 3.3). When sulphide was present, the direct exposure severely affected the initial phosphate uptake rates. At 42 mgTS-S/L, the initial phosphate uptake rate decreased 81% compared to the rate measured at 0 mgTS-S/L (13 mgPO$_4$-P /gVSS.h versus 68 mgPO$_4$-P /gVSS.h, respectively). It decreased even further and at 115 mgTS-S/L no phosphate uptake was observed. Furthermore, at 189 mgTS-S/L phosphate was not taken up but released at a rate of 12 mgPO$_4$-P/gVSS.h (Figure 3.4).

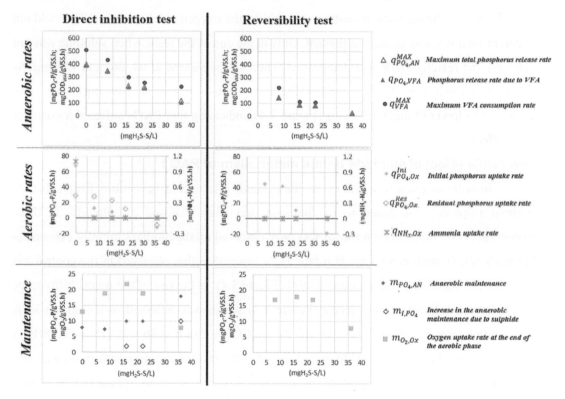

Figure 3.4.- Biomass specific anaerobic and aerobic rates during the direct inhibition tests and in the reversibility tests after the exposure at 42 mgTS-S/L (8 mgH$_2$S-S/L), 86 mgTS-S/L (16 mgH$_2$S-S/L), 115 mgTS-S/L (22 mgH$_2$S-S/L), and 189 mgTS-S/L (36 mgH$_2$S-S/L).

After sulphide was no longer detected (around 60 min), the aerobic P-uptake activities resumed (hereafter identified as residual phosphate uptake rate, $q_{PO_4,Ox}^{Res}$), reaching 28, 23 and 12 mg PO_4-P/gVSS.h at 48, 86 and 115 mg TS-S/L, respectively. Despite these uptake rates, suggesting that the sulphide inhibition effects were reversible, at 189 mgTS-S/L phosphate was released but a slower rate (9 mg PO_4-P/gVSS.h) than when sulphide was present (Figure 3.4).

Interestingly, in all direct inhibition tests ammonia consumption was not observed. Possibly, due to the chemical oxidation of sulphide, the oxygen uptake rate (OUR) could not be determined when sulphide was present. Only at the end of the tests, when sulphide was not observed, the residual OUR was determined. The OUR measured at 48, 86, and 115 mgTS-S/L were similar (19, 22, 19 mgO_2/gVSS.h, respectively) but at 189 mgTS-S/L was considerably lower (8 mgO_2/gVSS.h) (Figure 3.4), indicating that PAOs suffered a potential lethal effect.

Reversible effect of sulphide on the metabolism of PAO I

In order to assess to which extent the sulphide effects were reversible, fresh substrate (without sulphide) was added to the biomass previously exposed to sulphide. In these reversibility tests, full carbon removal was only observed on the biomass initially exposed to 42 mgTS-S/L (8 mgH$_2$S-S/L) but at a rate 57 % lower than that observed in the control test (510 and 221 mgCOD/gVSS.h, respectively) (Figure 3.4). In the rest of the reversibility tests, acetate leaked into the aerobic phase (Figure 3.3) which was reflected in maximum VFA consumption rates that dropped to 112 and 105 mgCOD/gVSS.h at 86 and 115 mgTS-S/L. A more pronounced effect by sulphide was observed at the reversibility test performed with 189 mgTS-S/L where almost all propionate and acetate leaked into the aerobic phase (24 mgCOD/L and 77 mgCOD/L respectively). The maximum anaerobic phosphate release rate ($q_{PO_4,AN}^{MAX}$) was also inhibited. It dropped from 399 to 144, 88, 81, and 25 mgPO4-P/gVSS.h as the sulphide concentrations in the direct inhibition test increased (conducted with 0, 42, 86, 115 and 189 mgTS-S/L) (Figure 3.4).

Moreover, the initial phosphate uptake rate ($q_{PO_4,Ox}^{Ini}$) decreased from 68, 45, 42, to 11 mg PO_4-P/gVSS.h in the reversibility tests with 0, 48, 86, and 115 mgTS-S/L (Figure 3.4). Similar to the observations in the aerobic phase of the direct inhibition test of 189 mgTS-S/L, in the reversibility test phosphate was not taken up, but released at a rate of 19 mgPO$_4$-P/gVSS.h.

Since in the reversibility test part of the carbon (VFA) leaked into the aerobic phase it was not possible to compare the initial OUR. Nevertheless, the $m_{O_2,Ox}$ measured in the reversibility test of 42, 86, and 115 mgTS-S/L were not considerably different between each other (17, 18, and 17 mgO$_2$/gVSS.h). Furthermore, similar to the findings of the direct inhibition test performed with 189 mg TS-S/L, in the reversibility test the $m_{O_2,Ox}$ dropped to 8 mgO$_2$/gVSS.h. Like in the direct inhibition tests, in the reversibility tests ammonia consumption was not observed.

Ions transport across the cell membrane in the direct and reversible inhibition batch test.

Phosphate is transported over the membrane together with counter ions like potassium, magnesium and calcium (Comeau et al., 1986). To assess if sulphide had an effect on the transport processes associated with P-release and uptake, the ratios between the phosphate and counter ions concentrations were assessed (Table 3.1). In the anaerobic phase, no considerable difference in the ratios of interest was observed. However, the ratios measured in the aerobic phase were more sensitive. At 86 and 115 mgTS-S/L, 0.06 and 0.89 mol-K$^+$ per P-mol, respectively, were released instead of stored (Table 3.1). The aerobic potassium-to-phosphate ratio was marginally restored in the reversible tests. In addition, above 86 mgTS-S/L a higher magnesium concentration was taken up together with phosphate (0.40 molMg^{2+}/P-mol at 115mgTS-S/L) but this ratio was fully restored to typical ratios in the reversible tests. In contrast, it was not possible to observe any transport of calcium in neither the anaerobic nor the aerobic phase of the batch test.

Table 3.1.- Molar ratios between counter ions and phosphate concentrations in the anaerobic and aerobic tests performed in this study.

			Direct inhibition tests mgTS-S/L				Reversibility tests mgTS-S/L		
Anaerobic ratios (release of phosphate and ions)									
Ratio	Comeau et al. (1986)	Smolders et al. (1994 a,b)	0	42	86	115	42	86	115
mol K$^+$/P-mol	0.20	0.33	0.29	0.32	0.32	0.36	0.27	0.26	0.25
mol Mg^{2+}/P-mol	0.28	0.33	0.27	0.32	0.31	0.33	0.37	0.37	0.36
mol Ca$^+$/P-mol	0.09	N.R.	0.01	-0.01	0.00	0.00	-0.01	0.00	-0.01
Aerobic ratios (uptake of phosphate and ions)									
mol K$^+$/P-mol	0.23	0.33	0.28	0.31	-0.06	-0.89	0.26	0.20	0.18
mol Mg^{2+}/P-mol	0.27	0.33	0.30	0.30	0.26	0.40	0.38	0.39	0.36
mol Ca$^+$/P-mol	0.12	N.R	0.00	0.05	-0.02	-0.06	0.09	-0.01	-0.03

*A negative sign means a switch of corresponding expected mechanism (e.g. release instead of taken up) N.R. Not reported

Mathematical description of the sulphide effects on PAO

The direct inhibition effects of sulphide on the anaerobic/aerobic phosphate profiles were satisfactorily described with the proposed expressions (NRMSD of 0.095 ± 0.006) (Figure 3.5). The increase in the net P-release was described as an increase in the anaerobic maintenance activity due to sulphide presence. The inhibition constant of sulphide ($K_{I,AN}$) for the anaerobic P-release was 20.7 mgTS-S/L and the maintenance coefficient ($m_{I,PO4}^{MAX}$) determined was 10 mgPO$_4$-P/gVSS.h (Equation 3.2).

To describe the sulphide effects on the aerobic physiology of PAO, the mathematical expressions considered that the effects were dependant on the actual concentrations of sulphide at the start of the aerobic phase ($S_{H_xS}^{Ini}$). In addition, 118 mgTS-S/L was used as the reference concentration at which a complete irreversible inhibition ($S_{H_xS}^{ref}$) occurred (Equation 3.4). In these expressions, an aerobic half- saturation inhibition constant ($K_{I,Ox}$) of 3.3 mgTS-S/L was used.

3.6. Discussion

Direct inhibition and reversibility effects of sulphide on the anaerobic metabolism of PAO I

At the direct inhibition test performed at 115 mgTS-S/L (22 mgH₂S-S/L) the VFA uptake rate decreased 50 % compared to the one observed in the control test. This H₂S concentration (which corresponds to about 22 mgH₂S-S/L) is lower than the 50 % inhibition concentration reported by Saad et al. (2013) of 60 mgH₂S-S/L. However, the experiments in this research were carried out at pH of 7.6, on the contrary Saad et al. (2013) performed their experiments within a pH range between 6.5 and 7.8. Thus, the difference in the sulphide inhibition observed in the past studies could be because of the pH used, as the external pH affects the transport of acetate and proton motor force (pmf) of PAO (Comeau et al., 1986; Smolders et al., 1994b).

In addition, the increase in the net P-released per VFA uptake from 0.78 to 0.91 P-mmol/C-mmol at 0 and 189mg TS-S/L, respectively, could be related to an increase in the anaerobic maintenance requirements ($m_{PO_4,AN}$) (Figure 3.4), which increases from 8 mgPO₄-P/gVSS.h to 18 mgPO₄-P/gVSS.h. These observations are in agreement with Comeau et al. (1986). They observed a similar increase and suggested that sulphide could disassociate inside the cell, reducing the internal pH (affecting the pH gradient of the cell), and increasing the P-release to stabilize the pH gradient across the cell membrane.

In the direct inhibition tests performed at 42, 86, and 115 mgTS-S/L, similar inhibition effects were observed on the maximum VFA uptake rate (q_{VFA}^{MAX}) and its associated P-release rate ($q_{PO_4,VFA}$) (of 85 % and 87 %, 58 % and 57 %, 50 % and 54 %, at the corresponding sulphide concentrations tested). Thus, this suggests that up to 115 mgTS-S/L, enough ATP (used for maintenance and acetate uptake) can be generated by poly-P hydrolysis. On the contrary, at 189 mgTS-S/L the maximum VFA uptake rate (q_{VFA}^{MAX}) and its associated P-release rate ($q_{PO_4,VFA}$) were of 66 % and 75 % lower than the rates observed in the control test. Hence, it seems that at 189 mgTS-S/L either (i) the ATP generated by poly-P hydrolysis is not enough, which could suggest a higher glycogen consumption or (ii) PAO has a lower ATP demand, which could be associated with a lethal effect of sulphide. Nevertheless, as glycogen was not measured, it is not possible to assess this hypothesis in this study.

The effect of sulphide on the VFA consumption seems to be more severely affected in the following reversible test as even at 86 mgTS-S/L an incomplete VFA consumption was observed. However, during the direct inhibition test not all the phosphorus previously released was aerobically taken up. Thus, the estimated initial poly-P content in each reversibility test decreased progressively from 0.13 to 0.12, 0.11 and 0.07 mgPO$_4$-P/mg VSS at the tests performed at 42, 86, 115 and 189 mgTS-S/L, respectively. Welles et al, (2015) demonstrated that the carbon uptake rate is affected by the poly-P content, which can also explain the slower carbon uptake rates seen in the reversibility tests.

Sulphide effects on the aerobic metabolism of PAO I

Based on the phosphorus profile, the aerobic metabolism seems to be more drastically affected than the anaerobic metabolism of PAO I (Figure 3.3). This agrees with past studies where the effect of different toxic compounds (e.g. free nitrous acid (FNA) or copper) inhibited more severely the aerobic metabolism of PAO (Wu et al., 2010; Chen et al., 2012; Fang et al., 2012).

The initial P-uptake ($q_{PO_4,Ox}^{Ini}$) and residual P-uptake rates ($q_{PO_4,Ox}^{res}$) decreased proportionally to the increase in the sulphide concentrations (Figure 3.4). The presence of sulphide could create stress conditions for PAOI, increasing the maintenance requirements and leaving less energy available for Poly-P formation. Furthermore, in the direct inhibition test performed at 189 mgTS-S/L, the energy (ATP) provided by the oxidation of PHA seems to have become limiting and the observed aerobic P-release was likely a consequence of the aerobic hydrolysis of poly-P for ATP generation. This statement is in agreement with Wu et al. (2010), Zhou et al. (2012) and Welles et al. (2015) who also observed aerobic P-release during the presence of FNA, copper or sodium chloride in EBPR cultures.

Furthermore, in the reversibility tests conducted at 48 and 86 mgTS-S/L, the initial phosphorus uptake rate was higher than the residual phosphate uptake rate observed in the direct inhibition tests (45, and 28 mgPO$_4$-P/gVSS.h and 42, and 23 mgPO$_4$-P/gVSS.h respectively), suggesting that the residual phosphate uptake rate ($q_{PO_4,0x}^{res}$) was only partially reversible up to 86 mgTS-S/L (Figure 3.4).

Ammonia uptake, which is usually associated with microbial growth, was not observed in neither the direct inhibition nor the reversibility tests. This observation is in agreement with Pijuan et al. (2010) and Welles et al. (2015) who observed that ammonia uptake was the most inhibited aerobic metabolic process by FNA and salinity, respectively. The lack of iron (which can precipitate with sulphide as FeS) has been suggested to affect the growth of aerobic microorganisms (Isa et al., 1986) and can be considered to be another potential inhibition mechanism. However, in these experiments iron was always present above the concentration of 0.4 mg Fe/L, which makes unlikely that the microbial growth (ammonia uptake) was inhibited due to iron limitation. Possibly, the inhibition of ammonia uptake might have been caused by the higher ATP growth requirements compared to the energy needs of other aerobic metabolic processes (e.g. 1.6 ATP/C-mol for growth, and 1.26 ATP/P-mol for PO$_4$-P uptake).

Due to the likely oxidation of sulphide in the direct inhibition tests and the VFA leakage in the aerobic stages of the reversibility tests, it was not possible to determine the OUR. Nevertheless, during the last minutes of the aerobic phases (around 2h) the OUR remained stable and it was assumed to correspond to certain aerobic maintenance or residual OUR ($m_{O_2,0x}$). In Figure 3.4, the $m_{O_2,0x}$ in the direct inhibition test increased from 13 to 22 mgO$_2$/gVSS.h at 0 and 86 mgTS-S/L, respectively, and thereafter decreased to 8 mgO$_2$/gVSS.h at 189 mgTS-S/L. Welles et al. (2015) observed a similar increasing effect on the aerobic maintenance energy requirements up to 2 % salinity followed by a decrease at 3 % salinity. Pijuan et al. (2010) determined that the anabolic processes of PAO (such as growth, glycogen replenishment and poly-P storage) were completely inhibited at $6 \cdot 10^{-3}$ mgHNO$_2$-N/L and the catabolic processes (maintenance) 40 to 50 % inhibited at 2 to $10 \cdot 10^{-3}$ mgHNO$_2$-N/L.

Sulphide effects on the transport of cations across the cell membrane

Similar to the transport of phosphate, the aerobic transport of cations (e.g. calcium, magnesium, and potassium) across the cell membrane was more affected than the anaerobic transport (Table 3.1). In agreement with Pattarkine et al. (1999) it seems that calcium was not used for the transport of phosphate across the cell membrane. Sulphide affected marginally the Mg^{2+}/P ratio at the sulphide concentrations tested. However, in the direct inhibition test potassium was released above 86 mg TS-S/L. As suggested elsewhere, both potassium and magnesium are essential for P-uptake, and the absence of one of them can result in the failure of the EBPR system once poly-P is depleted (Rickard et al., 1992; Brdjanovic et al., 1996; Pattarkine et al., 1999; Barat et al., 2005). Nevertheless, since in the reversibility tests potassium was taken up together with phosphate (Table 3.1), a shock of sulphide (up to 16 mgH$_2$S-S/L) would not likely lead to EBPR failure due to poly-P depletion.

Furthermore, as K$^+$ and Mg^{2+} are co-transported across the cell membrane together with phosphate, the K-release could have implied that phosphate was also released. However, excluding the direct inhibition and reversibility tests performed at 189 mgTS-S/L, aerobic P-release was not observed. As suggested by Comeau et al. (1986) potassium and magnesium were likely released to re-establish the pH gradient across the cell membrane, which can be supported by the higher mol K$^+$/P-mol ratio that increased from 0.29 to 0.36 mol K$^+$/P-mol in the direct inhibition test performed at 0 and 115 mgTS-S/L.

Limitations and applications of the mathematical description of the effects of sulphide on PAO I

The equations used to model the phosphate profile of these experiments, are based in the assumption that sulphide affect the metabolism of PAO I due to diffusion into the cell membrane. The diffusion of sulphide into the cell membrane can cause an increase in the energy requirements, which results in a higher anaerobic maintenance and slower phosphate uptake rate. Hence, the mathematical expression proposed could be used to try to predict the total anaerobic phosphate release and phosphate uptake rate of an enriched culture of PAO I up to 189 mgTS-S/L. For example, the operator of a wastewater treatment plant (WWTP) could measure the concentration of sulphide at the start of the anaerobic and aerobic tanks, and use these mathematical expressions to increase the aerobic retention time or to identify the chemical dose that would need to be added to meet the phosphate effluent standard. Nevertheless, these mathematical equations do not include pH, which affect the speciation of sulphide (HS^-+H_2S). Moreover, the presence of other microorganism capable to oxidize sulphide might reduce the inhibition effect of sulphide on PAO I.

Sulphide effects on full scale EBPR systems

These experiments focus on the short-term (hours) effects during and after the exposure of PAO to sulphide. Such conditions can occur in sewage treatment plants that regularly receive saline wastewater but also in WWTP exposed to the sudden discharge of industrial effluents rich in sulphate or WWTP in coastal zones subject to saline intrusion into the sewer due to tidal events or poor conditions of sewage pipes. Such conditions may lead to process upsets and deterioration if the sulphide concentrations exceed 8 mg H_2S-S/L, as potentially the growth of PAO could be inhibited. However, the long-term exposure to sulphide could lead to biomass acclimatization or selection of sulphide-tolerant PAO species or sulphide oxidizing organism (Schulz et al., 1999; Brock et al., 2012; Ginestet et al., 2015). These organisms could reduce the deleterious effects of sulphide on EBPR systems. Further research is needed to assess the long-term effects of sulphide on EBPR systems.

3.7. Conclusions

In the present study, sulphide affected more severely the aerobic metabolism than the anaerobic metabolism of PAO. The I_{50} for the VFA uptake rate (q_{VFA}^{MAX}) and phosphate release rate ($q_{PO_4,AN}^{MAX}$) were estimated to be around 115 mgTS-S/L (22 mgH$_2$S-S/L). The effect of sulphide on the aerobic P-uptake rate was partially reversible up to 115 mgTS-S/L. Phosphate and potassium were released when the concentrations were above 189 and 86 mgTS-S/L, respectively, indicating that PAO experienced a strong inhibition by sulphide. Ammonia uptake was not observed in neither the direct nor the reversible tests, suggesting that PAO was unable to grow when sulphide was present.

4

Long-term effects of sulphide on the enhanced biological removal of phosphorus: The symbiotic role of *Thiothrix caldifontis*

Chapter 4
Long-term effects of sulphide on the enhanced biological removal of phosphorus:
The symbiotic role of *Thiothrix caldifontis*

4.1. Highlights

- Satisfactory EBPR was achieved up to 100 mg H2S-S/L in influent.

- *Thiothrix caldifontis* played a major role through mixotrophic growth.

- *Thiothrix caldifontis* stored acetate as PHA during anaerobic conditions.

- *Thiothrix caldifontis* could store up to 100 mg P/gVSS during aerobic conditions.

- Polysulfide oxidation appeared to provide the required energy under aerobic conditions

Adapted from

Rubio-Rincón F.J., Welles L, Lopez-Vazquez C.M., Nierychlo M., Abbas B, Geleijnse M., Nielsen P.H., van Loosdrecht M.C.M., Brdjanovic D. (submitted) Long-term effects of sulphide on the enhanced biological removal of phosphorus: The symbiotic role of *Thiothrix caldifontis*.

Chapter 4
Long-term effects of sulphide on the enhanced biological removal of phosphorus:
The symbiotic role of *Thiothrix caldifontis*

4.2.   Graphical Abstract

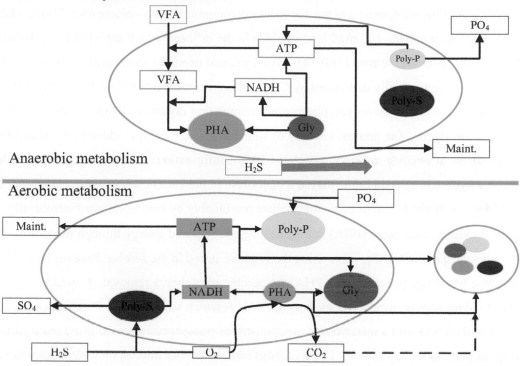

Graphical abstract.- Proposed anaerobic and aerobic metabolism of *Thiothrix caldifontis* .

Chapter 4
Long-term effects of sulphide on the enhanced biological removal of phosphorus:
The symbiotic role of *Thiothrix caldifontis*

4.3. Abstract

Thiothrix caldifontis was the dominant microorganism (bio-volume 65 ± 3 %) in a lab-scale system containing 100 mgS/L of sulphide in the influent, which achieved full enhanced biological phosphorus removal (EBPR). After a gradual exposure to the presence of sulphide, the EBPR system initially dominated by *Candidatus* Accumulibacter phosphatis Clade I (bio-volume 98 ± 3 %) (a known polyphosphate-accumulating organism, PAO) became enriched with *T. caldifontis*. The gradual increase of sulphide in the medium (added to the anaerobic stage of an alternating anaerobic-aerobic EBPR configuration) and the adjustment of the aerobic hydraulic retention time played a major role in the enrichment of these organisms. *T. caldifontis* exhibited a mixotrophic metabolism presumably by storing carbon anaerobically as poly-β-hydroxyalkanoates (PHA) and generating the required energy through the hydrolysis of poly-phosphate (Poly-P), while using the sulphide stored in the aerobic stage (as Poly-S) as a source of energy (together with PHA) for aerobic phosphorus removal. *T. caldifontis* was able to store approximately 100 mgP/gVSS. This research suggests that *T. caldifontis* could behave like PAO with a mixotrophic metabolism for phosphorus removal using intracellular sulphur pool (as energy source). These findings can be of major interest for biological removal of phosphorus from wastewaters with low organic carbon concentrations containing reduced S-compounds, like those (pre-) treated in anaerobic systems or after anaerobic sewer systems.

Keywords: *Thiothrix caldifontis, Candidatus* Accumulibacter, autotrophic phosphorus removal, biological removal of phosphorus, sulphide oxidation.

Chapter 4
Long-term effects of sulphide on the enhanced biological removal of phosphorus:
The symbiotic role of *Thiothrix caldifontis*

4.3. Introduction

The enhanced biological phosphorus removal (EBPR) process is broadly applied in sewage treatment plants to meet the phosphorus discharge standards of treated wastewater. In this process, phosphorus is removed by polyphosphate accumulating organisms (PAOs) that store phosphorus beyond their growth requirements and are enriched by recirculating the activated sludge mixed liquor through anaerobic and aerobic/anoxic conditions (Barnard, 1975). Under anaerobic conditions, PAOs store volatile fatty acids (VFAs) present in the wastewater influent as poly-β-hydroxy-alkanoates (PHAs), using the energy generated from the hydrolysis of poly-phosphate (Poly-P) and glycogen. In the subsequent aerobic/anoxic phases, PAOs oxidize the stored PHA to restore their Poly-P storage pools (resulting in the biological removal of phosphorus from the water phase) as well as to replenish the intracellular glycogen pools, for biomass synthesis and maintenance purposes (Comeau et al., 1986; Mino et al., 1998). A member of the family *Rhodocyclaceae* (identified as genus "*Candidatus* Accumulibacter") has been suggested to be one of the main PAO involved in the EBPR process in wastewater treatment plants (WWTPs) (Hesselmann et al., 1999; Seviour et al., 2003b).

In urban environments, wastewaters with sulphate concentrations of up to 500 mg SO_4^{2-}/L can be formed due to: (i) discharge of industrial effluents rich in sulphate (Sears et al., 2004), (ii) use of sulphate-based chemicals in drinking water treatment (e.g. aluminium sulphate) (Bratby, 2016), (iii) saline (sea and brackish) water intrusion into the sewage network, and (iv) the direct use of sea and brackish water as secondary quality water (e.g. cooling, toilet flushing) (Lee et al., 1997). In the absence of other electron acceptors, sulphate can be reduced to sulphide (Koster et al., 1986). At concentrations as low as 8 mg H_2S-S/L, sulphide has been observed to inhibit the anaerobic and (more severely) the aerobic metabolism of *Ca.* Accumulibacter after a sudden short-term exposure to this compound (Comeau et al., 1986; Rubio-Rincon et al., chapter 3). Yamamoto et al. (1991) and Baetens et al. (2001) assessed the long-term effects of the sulphate reduction process (resulting in H_2S formation) on the EBPR process, but the proliferation of filamentous bacteria led to the failure of their

Chapter 4
Long-term effects of sulphide on the enhanced biological removal of phosphorus:
The symbiotic role of *Thiothrix caldifontis*

systems at concentrations even lower than those studied by Rubio-Rincon et al. (2016).

Interestingly, some bacteria have the ability to use sulphide as energy source for the intracellular accumulation of phosphorus (Schulz et al., 2005; Brock et al., 2011; Ginestet et al., 2015; Guo et al., 2016b). Schulz et al. (2005) observed that under anaerobic conditions *Thiomargarita namibiensis* used their intracellularly stored nitrate and phosphorus to oxidize sulphide into sulphur and stored it as poly-sulphur (Poly-S). While acetate triggered the anaerobic metabolism, PHA inclusions were not observed. Instead, acetate was stored as glycogen. When an electron acceptor was available, *T. namibiensis* generated the required energy from Poly-S and glycogen to replenish their Poly-P storage pools (Schuler, 2005). Also, Brock et al. (2011) observed that a marine *Beggiatoa* strain was capable to store Poly-P above their growth requirements using intracellularly stored Poly-S as a source of energy. But, contrary to the metabolism of *Ca.* Accumulibacter, the anaerobic phosphorus released was not affected by the addition of VFA. The phosphorus release observed under anaerobic conditions was only associated to maintenance requirements, which increased proportionally to the concentration of sulphide.

Ginestet et al. (2015) observed that certain organisms could use intracellularly stored sulphur to generate energy for phosphorus uptake in a full-scale wastewater treatment plant. However, the microbial communities were not identified and the phosphorus uptake rates and net phosphorus removal were relatively low (2.9 mg PO_4-P/L). Guo et al. (2016) observed that the addition of sulphide improved the biological removal of phosphorus, where sulphur oxidizing bacteria (SOB) were involved in the main EBPR process. However, the SOB culture comprised only 2.6% of the total biomass with a net P-removal of 27 mg P/gVSS, that, arguably, is close to the phosphorus growth requirements of conventional systems (estimated around 20 to 30 mg P/gVSS) (Henze et al., 2008). One of the most common sulphur oxidizing bacteria observed in EBPR belong to the genera of *Thiothrix* (Wanner et al., 1987; García Martín et al., 2006; Gonzalez-Gil et al., 2011). Some strains from this genera, as *Thiothrix caldifontis* had been isolated and studied in the past, where it has been show that the bacteria

74

Chapter 4
Long-term effects of sulphide on the enhanced biological removal of phosphorus:
The symbiotic role of *Thiothrix caldifontis*

is capable to aerobically grow mixotrophically and store sulphide or thiosulphate as Poly-S for its later oxidation into sulphate (Chernousova et al., 2009). Nevertheless, for the best of our knowledge there is not literature which could indicate that this bacterium is capable to store phosphorus beyond its growth requirements.

Due to the increasing generation of saline wastewaters rich in sulphate and the commitment to reduce the phosphorus emissions, there is a need to assess the long-term effects of sulphide on the EBPR process. This will allow to addressing any potential selection or adaptation of sulphide tolerant PAO species or side populations that can contribute to reduce the deleterious effects of sulphide, while achieving satisfactory biological P-removal. For this purpose, over a year, a lab-scale anaerobic-aerobic EBPR system, initially enriched with *Ca.* Accumulibacter, was exposed to different sulphide concentrations fed with the medium to the anaerobic stage.

4.4. Materials and methods

Reactor operation

Prior to the execution of this study, a 3.0 L Applikon reactor with double-jacket and working volume of 2.5 L (Delft, Netherlands) enriched with *Candidatus* Accumulibacter phosphatis clade I was operated for more than 200 days showing a stable EBPR performance (Rubio-Rincon et al., chapter 3). When the present study started, 50 mL of activated sludge from the WWTP Nieuwe Waterweg (Hoek van Holland, The Netherlands) were added and two experimental phases were carried out. In the first phase, a cycle of 8h was applied (composed of 2h 15 min anaerobic, 4h aerobic, 1h 15 min settling and 30 min effluent removal stages). Due to phosphorus was not aerobically completely taken up at 20 mgS/L, the aerobic phase was increased to 5h and the anaerobic phase was reduced to 1h 15 min in order to keep the same cycle length (experimental phase two) At the start of each cycle nitrogen gas was sparged for 15 min (at a flowrate of 70 L/h) to create anaerobic conditions and 1.25L of synthetic media

Chapter 4
Long-term effects of sulphide on the enhanced biological removal of phosphorus:
The symbiotic role of *Thiothrix caldifontis*

were fed in the next 5 min. Thereafter, sulphide was fed at different concentrations gradually increased from 10 to 100 mgS/L. In order to control the solids retention time (SRT) at 20 d, 41 mL of mixed liquor were withdrawn at the end of each aerobic phase. The pH was kept at 7.6 ± 0.1 through the addition of 0.1M HCl and 0.4M NaOH. In the aerobic stages, compressed air was supplied at a flowrate of 20 L/h. Temperature was controlled at 20±1°C. Volatile fatty acids (VFA), ortho-phosphate (PO_4-P), sulphide, volatile suspended solids (VSS), and total suspended solids (TSS) concentrations were measured twice per week at the start and end of each phase. When no significant changes in the concentrations of these parameters were observed for at least 30 days, it was considered that the system had reached pseudo steady-state conditions.

Media

The synthetic media fed to the reactor contained per litre 637 mg NaOAc•3H2O (295 mgCOD/L), 66.7 μL propionic acid (100 mg COD/L), 107 mg NH_4Cl, 135 mg NaH_2PO_4•H_2O (30 mgPO_4-P/L), 90 mg $MgSO_4$•$7H_2O$, 14 mg $CaCl_2$•$2H_2O$, 36 mg KCl, 1 mg yeast extract, 20 mg N-allylthiourea (ATU) and 300 μL of trace element solution prepared according to Smolders et al. (1994). Sulphide was fed from a bottle with a stock solution containing 3.2 gS/L and 0.1 M NaOH. When the sulphide solution was added to the reactor, a 0.1 M HCl solution was fed simultaneously to keep the pH at the desired set point (7.6 ± 0.1).

Cycle Tests

When the activity of the parent reactor reached pseudo steady-state conditions, cycle tests were conducted in the parent reactor (3.0 L). Samples for the determination of VFA, ortho-phosphate, sulphide, and ammonia concentrations were taken along the cycle test; while samples for the determination of PHA, glycogen, VSS, TSS, magnesium, calcium, and potassium were collected at the start and end of each phase.

Chapter 4
Long-term effects of sulphide on the enhanced biological removal of phosphorus:
The symbiotic role of *Thiothrix caldifontis*

Batch Activity Tests

Once the gradual addition of sulphide reached 100 mg TS/L and the parent reactor exhibited a pseudo steady-state activity, additional anaerobic-aerobic (1h and 3h, respectively) batch activity tests were conducted to assess the physiology of the enriched culture. Five batch activity tests were executed with or without VFA (acetate), sulphide and/or biomass (Table 4.1). Ortho-phosphate, sulphide, sulphate, and ammonia were measured at different times along the batch test. VSS and TSS were measured at the start, end of the anaerobic and end of aerobic phase.

Table 4.1.- Batch activity tests conducted to assess the physiology of the enriched EBPR culture.

Batch test	VFA mg COD/L	Sulphide mg S/L	mg VSS	Target activity or purpose
1	100	45	540	Simultaneous PAO and sulphur oxidation activities
2	100	None	595	PAO activity
3	None	45	877	Sulphide oxidation process
4	None	None	823	Blank
5	None	45	None	Control test and potential chemical sulphide oxidation

Analytical Measurements

Ortho-phosphate (PO_4^{3-}-P), ammonium (NH_4^+-N), sulphide (H_2S+HS^-), VSS and TSS were measured as described in APHA et al. (2005). Sulphate was measured using an Ion Chromatography system equipped with a Dionex Ionpack AS4A-SC column (Dreieich, Germany). Potassium (K^+), magnesium (Mg^{2+}) and calcium (Ca^+) were measured in an Inductively Coupled Plasma, Mass Spectroscopy manufactured by Thermo Scientific in Bremen, Germany. Acetate (HAc) and propionate (HPr) were measured in a gas chromatography system G420-C (Nieuwegein, The Netherlands). All analyses were performed

Chapter 4
Long-term effects of sulphide on the enhanced biological removal of phosphorus:
The symbiotic role of *Thiothrix caldifontis*

within 2 hours after the cycle test finished and handled as described elsewhere (Rubio-Rincon et al., chapter 3).

Kinetic and stoichiometric parameters of interest

The net P-released-to-VFA consumption (P-mol/C-mol), net PHA stored-to-VFA consumption (C-mol/C-mol) and the net glycogen utilization-to-VFA consumption ratio (C-mol/C-mol) were the anaerobic stoichiometric parameters of interest. Also, the potassium-to-phosphorus (mol-K^+/P-mol), magnesium-to-phosphorus (mol-Mg^{2+}/P-mol), and calcium-to-phosphorus (mol-Ca^+ /P-mol) conversion ratios in the anaerobic and aerobic stages were calculated. The Poly-P content of the biomass and observed growth were estimated with a mass balance as described by Kuba et al. (1993). The kinetic rates were calculated by linear regression as described in Smolders et al. (1995). The kinetic rates of interest were:

q_{VFA}^{MAX}:	Maximum specific VFA consumption rate, in mg COD/gVSS.h
$q_{PO_4,AN}^{MAX}$:	Maximum specific total phosphorus release rate, in mg PO_4-P/gVSS.h
$q_{PO_4,Ox}$:	Specific phosphorus uptake rate, in mg PO_4-P/gVSS.h
$q_{NH_x,Ox}$:	Specific ammonia uptake rate, in mg NH_4-N/gVSS.h
$q_{H_2S,Ox}$:	Specific sulphide oxidation rate, in mg S/gVSS.h
$q_{O_2,Ox}$:	Oxygen consumption rate, in mg O_2/gVSS.h

Microbiological characterization and identification

Microscopic identification of intracellular polymers

To preserve the sludge samples, 6 drops of aldehyde were added to 10 mL of sludge samples immediately after collection as described in Marzluf et al. (2007). PHA inclusions in the biomass were detected using Nile Blue A and BODIPY 505/515 applied according to Seviour et al. (2010) and Cooper et al. (2010), respectively. Poly-P was identified using DAPI as described in Seviour et al. (2010). Images were collected with an Olympus BX5i microscope

Chapter 4
Long-term effects of sulphide on the enhanced biological removal of phosphorus:
The symbiotic role of *Thiothrix caldifontis*

equipped with a SC100 and XM10 cameras (Hamburg, Germany).

Fluorescence *in situ* Hybridization (FISH)

In order to identify the most representative microbial communities, Fluorescence *in situ* Hybridization (FISH) analyses were performed according to Amman (1995). PAO were targeted with the PAOMIX probe (composed of probes PAO 462, PAO 651 and PAO846) (Crocetti et al., 2000). The presence of PAO clade I and clade II was determined through the addition of probes Acc-1-444 (1A) and Acc-2-444 (2A, 2C, 2D) (Flowers et al., 2009). *Candidatus Competibacter phosphatis*, a known glycogen-accumulating organism, was targeted with the GB probe (Kong et al., 2002). Defluvicoccus cluster 1 and 2 were identified with the TFO-DF215, TFO-DF618, DF988, and DF1020 probes (Wong et al., 2004; Meyer et al., 2006). In order to target filamentous bacteria from the Thiothrix genera, the probe G123T was used as described by Kanagawa et al. (2000). Vectashield containing a DAPI concentration was used to amplify the fluorescent signal and stain living organisms (Nielsen et al., 2009). The estimation of the relative abundance of the organisms of interest was performed as described by Flowers et al. (2009).

MAR-FISH (Microautoradiography-Fluorescence *in situ* Hybridization)

To address the potential anaerobic carbon uptake of the biomass, samples were collected from the reactor at the end of the anaerobic and aerobic phases and cooled down at 4°C. MAR incubations were performed within 24 h after sampling as described previously (Nielsen et al., 2005). To exhaust the PHA contents of the cells while maintaining their intracellular sulphur storage pools, biomass sample taken at the end of the anaerobic phase was pre-incubated aerobically at room temperature for 4 h in a 5 mM thiosulfate solution. Sample collected at the end of the aerobic phase (containing neither PHA nor S granules) were pre-incubated anaerobically at room temperature for 1 h for acclimatization purposes and to remove any potential residual concentration of an electron acceptor or donor. Thereafter, both

Chapter 4
Long-term effects of sulphide on the enhanced biological removal of phosphorus:
The symbiotic role of *Thiothrix caldifontis*

samples were diluted to 1gSS/L biomass concentration in a synthetic media similar to that fed to the parent reactor but excluding any carbon source. Two mL of sample was introduced to 10 mL serum bottles and sealed with thick rubber stoppers. To achieve anaerobic conditions, oxygen was removed before substrate addition through the repeated evacuation of headspace and injection of oxygen-free nitrogen. Unlabeled and tritium-labeled acetate (PerkinElmer Inc., Waltham, MA, USA) were degassed and added to serum vials under a nitrogen gas flow to reach final concentrations of 6 mM and 30 µCi/mL, respectively. After 3 h incubation at room temperature, cells were immediately fixed with cold 4% [w/v] paraformaldehyde (final concentration), washed three times with sterile filtered tap water and re-suspended in 1 ml of 1:1 PBS/EtOH solution. Aliquots of 30 µL were applied onto coverslips and the FISH procedure using EUBmix targeting most bacteria (Amann et al., 1990; Daims et al., 1999), labeled with 5(6)-carboxyfluorescein-N-hydroxysuccinimide ester (FLUOS) and G123T targeting *Thiothrix* (Kanagawa et al., 2000) labeled with cyanine Cy3 was performed as described previously. Thereafter, microscope slides were coated with Ilford K5D emulsion (Polysciences Inc., Warrington, PA, USA), exposed in the dark for 10 days at 4°C and revealed with Kodak D-19 developer (Artcraft Camera and Digital, Kingston, NY, USA). Microscopic analyses were carried out with an Axioskop epifluorescence microscope (Carl Zeiss, Oberkochen, Germany).

Community composition analysis using Denaturing Gradient Gel Electrophoresis (DGGE) and 16S rRNA Amplicon Sequencing

In order to confirm the results gathered using the FISH technique, and to further identify the species of microorganism present in our biomass, 16S-rDNA-PCR DGGE was applied. The complete procedure was carried out as described by Bassin et al. (2011). Additionally, community composition in the reactor fed with 100 mgS/L was analysed using 16S rRNA amplicon sequencing. DNA was extracted using the FastDNA® Spin kit for soil (MP Biomedicals, USA), V1-3 variable region of the 16S rRNA gene was amplified and

80

Chapter 4
Long-term effects of sulphide on the enhanced biological removal of phosphorus:
The symbiotic role of *Thiothrix caldifontis*

samples were prepared and sequenced as described in Albertsen *et al.*, (2015). The sequences were classified using MiDAS taxonomy, v. 1.20 (McIlroy et al., 2015). All sequenced sample libraries were subsampled to 10,000 reads and data analysis was performed in R v. 3.2.3 (R Core Team, 2015) using ampvis package v. 1.24.0 (Albertsen et al., 2015).

4.5. Results

Effect on the process performance of the biological removal of phosphorus by increasing sulphide concentrations

Prior to the beginning of this study, the parent SBR was operated with a SRT of 8 days for more than 200 days (data not shown). In the anaerobic phase, all VFA were taken up at a specific rate of 534 mg COD/gVSS.h, releasing 377 mg PO$_4$-P/gVSS.h. Under aerobic conditions, ortho-phosphate was removed at 57.9 mg PO$_4$-P/gVSS.h. According to FISH image analysis, *Ca.* Accumulibacter clade I was the dominant microorganism in the system (reaching a relative abundance of 99% with respect to DAPI) (Rubio-Rincon et al., 2016).

In the first experimental phase, the extension of the SRT to 20 d and the addition of 10 mgS/L slightly decreased the anaerobic and aerobic rates with regard to those observed at the 8 d SRT when no sulphide was added (Table 4.2). The marginal inhibition of the VFA uptake rate continued up to 20 mgS/L (from 370 to 194 mg COD/gVSS.h at 10 and 20 mgS/L, respectively). In a similar manner, the phosphorus release rate ($q_{PO_4,AN}^{MAX}$) decreased from 257 to 187 mg PO$_4$-P/gVSS.h. However, at 20 mg S/L not all phosphorus was aerobically taken up. Therefore, the aerobic phase was extended to 5 h (experimental phase 2), while the overall hydraulic retention time (HRT) and the sulphide concentration were kept at 18 h and at 20 mgS/L, respectively.

Chapter 4
Long-term effects of sulphide on the enhanced biological removal of phosphorus:
The symbiotic role of *Thiothrix caldifontis*

Once the EBPR system was stable, the sulphide concentrations were increased, but above 20 mgS/L the VFA uptake rate remained around 202±15 mg COD/gVSS.h. On the contrary, the maximum phosphorus release ($q_{PO_4,AN}^{MAX}$) decreased from 174 to 124 mg PO$_4$-P/gVSS.h as the sulphide addition increased from 20 to 100 mgS/L. The P/VFA ratio remained at 0.73±0.02 P-mol/C-mol up to 50 mgS/L, but decreased to 0.64 P-mol/C-mol at 100 mgS/L. A lower Poly-P content and a higher GLY/VFA ratio were observed as sulphide reached 100 mgS/L (Table 4.2). However, the PHA/VFA decreased from 1.00 to 0.76 C-mol/C-mol as the sulphide concentrations reached 100 mgS/L. The anaerobic COD balance closed to 99% and to 81% at 10 mg S/L and 100 mg S/L, respectively. In the anaerobic stages, the sulphide and ammonia concentrations remained relatively stable (Annex 4.A).

Regarding the aerobic metabolism, 20 mgS/L inhibited the aerobic phosphorus uptake, not all phosphorus was removed and 5 mg PO$_4$-P/L remained in the effluent (Annex 4.A) (Table 4.2). As previously described, this led to the start of the second experimental phase where the anaerobic phase was shortened from 2 h 15 min to 1 h 15 min and the aerobic stage was extended from 4 h to 5 h. Extending the aerobic phase helped to gradually increase the net aerobic phosphorus uptake from 27.6 mg PO$_4$-P/gVSS.h at 20 mgS/L (when phosphorus was observed in the effluent) to 61.1 mg PO$_4$-P/gVSS.h at 100 mgS/L, contributing to achieve full P-removal (Figure 4.1B). Moreover, after the biomass acclimatization and extension of the aerobic stage, the EBPR activity observed at 100 mgS/L (Figure 1B) was even higher than that observed with 10 mgS/L (Figure 4.1A). At 100 mgS/L the sludge did not settle correctly resulting in the loss of biomass through the effluent (around 220 mg TSS/cycle), leading to a reduction of the SRT from 20 d to approx. 4.5 d, which remained until the end of the present study.

Chapter 4
Long-term effects of sulphide on the enhanced biological removal of phosphorus:
The symbiotic role of *Thiothrix caldifontis*

Table 4.2.- Anaerobic and aerobic kinetic rates and stoichiometric ratios observed in the cycle tests as a function of the sulphide concentrations tested (from 10 to 100 mgS/L).

Sulphide mgS/L	q_{VFA}^{MAX}	$q_{PO_4,AN}^{MAX}$	P/VFA	PHV/VFA	PHB/VFA	PHA/VFA	Gly/PHA
	mg COD/ gVSS.h	mg PO$_4$-P/ gVSS.h	P-mmol/ C-mmol	C-mmol/ C-mmol	C-mmol/ C-mmol	C-mmol/ C-mmol	C-mmol/ C-mmol
10	370.4	257.0	0.75	0.31	0.69	1.00	0.12
20	268.5	187.4	0.75	N.M	N.M	N.M	N.M
20[a]	193.6	174.1	0.74	N.M	N.M	N.M	N.M
30[a]	213.0	198.0	0.72	N.M	N.M	N.M	N.M
50[a]	218.0	175.0	0.71	N.M	N.M	N.M	N.M
100[a]	187.0	124.0	0.64	0.21	0.54	0.76	0.20

Aerobic kinetic and stochiometric rates

	$q_{PO_4,Ox}$	$q_{NH_x,Ox}$	$q_{H_2S,Ox}$	$q_{O_2,Ox}$	Mg/P	K/P	Poly-P[b]
	mg PO$_4$-P/ gVSS.h	µg NH$_4$-N/ gVSS.h	mg H$_2$S-S/ gVSS.h	mg O$_2$ / gVSS.h	Mg-mmol/ P-mmol	K-mmol/ P-mmol	mgP/ mg VSS
10	43.5	593.7	13.0	19.3	N.M	N.M	0.41
20	27.6	432.9	23.1	17.7	N.M	N.M	0.25
20[a]	36.1	178.8	34.4	15.1	N.M	N.M	0.36
30[a]	56.4	884.6	131.4	23.7	0.33	0.36	N.C
50[a]	51.0	677.8	140.0	36.2	0.30	0.28	N.C
100[a]	61.1	1290.6	333.3	34.9	0.33	0.35	0.19

[a] Experiments performed according to experimental phase 2 (5h aerobic phase)
[b] Estimated as described in Kuba et al. (1993)
[c] N.M. Not measured; N.C. Not calculated

Chapter 4
Long-term effects of sulphide on the enhanced biological removal of phosphorus:
The symbiotic role of *Thiothrix caldifontis*

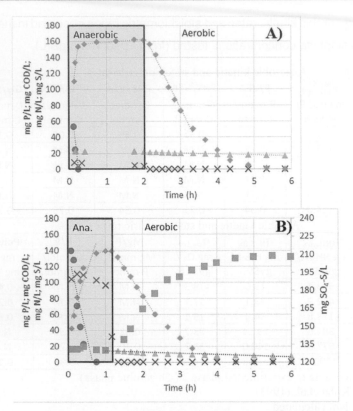

Figure 4.1.- VFA (circle), phosphorus (diamond), ammonia (triangle), sulphide (cross) and sulphate (square) profiles of the cycle tests performed at (A) 10 mgS/L with an aerobic phase of 4h and (B) 100 mgS/L with a duration of the aerobic stage of 5h.

The Mg/P and K/P ratios measured in the tests were in line with the theoretical composition of Poly-P ($Mg_{1/3}K_{1/3}PO_3$; Smolders et al., 1994), contributing to discard the potential chemical precipitation of phosphorus. In addition, no considerable changes in the concentrations of calcium were detected (data not shown). Nevertheless, a higher consumption of ammonia was observed which increased from 593.7 up to 1290.6 μgNH_4-N/gVSS.h at 10 and 100 mg S/L, respectively. Simultaneously, the oxidation of sulphide reached up to 333.3 mgS/gVSS.h at 100 mgS/L, whereas at 10 mgS/L had been only 13.0 mgS/gVSS.h (Figure

Chapter 4
Long-term effects of sulphide on the enhanced biological removal of phosphorus:
The symbiotic role of *Thiothrix caldifontis*

4.1). It was observed that the production of sulphate was slower than the oxidation of sulphide (39.4 mg SO_4-S/gVSS.h at 100 mg S/L). Moreover, only around 58 % of the sulphide fed to the system was oxidized to sulphate, suggesting that either there were certain intermediate compounds formed or that certain sulphur compounds were stored intracellularly.

Putative role of sulphide in the biological removal of phosphorus

In order to assess the potential effects of sulphide on the biological removal of phosphorus, five batch tests were performed (Table 4.1) with biomass from the system when 100mgS/L were fed. Figure 4.2 shows the phosphorus (Figures 4.2A and 4.2B), ammonia (Figure 4.2C), and sulphide and sulphate (Figure 4.2D) profiles observed in the different batch tests. In the batch tests fed with VFA (Figure 4.2A; diamond and square markers), the net phosphorus removal values were comparable between each other regardless whether sulphide was or was not added (96 and 92 mg PO_4-P/gVSS, respectively). On the other hand, in the tests where no VFA were added (Figure 4.2B; triangle and circle markers), the phosphorus removed was almost 2-fold higher in the experiment carried out with sulphide than without sulphide (13 and 6.1 mg PO_4-P/gVSS, respectively). Interestingly, the aerobic ammonia consumption was also two-fold higher in the experiments conducted with sulphide independently of whether VFA were dosed or not. Moreover, in none of the batch test neither nitrate nor nitrite were detected during the aerobic stage and no ammonia consumption was observed in the anaerobic stage. In addition, in the aerobic stages, a complete oxidation of sulphide was observed within the first 10 min in the control experiment (with biomass present); whereas in the blank test (performed without biomass), the oxidation of sulphide took two hours. Moreover, the sulphate production was slower than the sulphide consumption. These observations indicate that sulphide was biologically oxidized and presumably stored intracellularly as elemental sulphur in the first minutes of the aerobic stage (likely as Poly-S) and afterwards oxidized and released as sulphate.

Chapter 4
Long-term effects of sulphide on the enhanced biological removal of phosphorus:
The symbiotic role of *Thiothrix caldifontis*

Selection and adaptation of the microbial community

In view of the stable EBPR activity at relatively high sulphide concentrations (100 mgS/L), different microbiological characterization and identification analyses were performed to assess the potential selection or adaptation of sulphide-tolerant PAOs. FISH image analyses indicated that *Ca.* Accumulibacter (PAO clade I) were the dominant microorganisms at 10 mgS/L (Figure 4.3A) comprising around 76±2% of the total bacterial population, while filamentous bacteria composed of only 4±0.5%. However, at 100 mgS/L, the fraction of filamentous organisms increased up to 65±3%, while that of *Ca.* Accumulibacter decreased to 33±2% (Figure 4.3; Annex 4.B). Using 16S rRNA amplicon sequencing, a relative abundance of 81% of OUT's belonging to *Thiothrix* genus was obtained (Figure 4.4). DGGE analyses supported these observations, indicating an increase in the dominance of *T. caldifontis* (99 % similarity) in the reactor when it was fed with 100 mgS/L (Figure 4.5, bands 11, and 21). Based on *ppk* analysis, the main *Ca.* Accumulibacter clade switched from IC to IA, as the sulphide concentration increased (Figure 4.6). This could indicate the selection of a sulphide tolerant *Ca.* Accumulibacter culture.

Chapter 4
Long-term effects of sulphide on the enhanced biological removal of phosphorus:
The symbiotic role of *Thiothrix caldifontis*

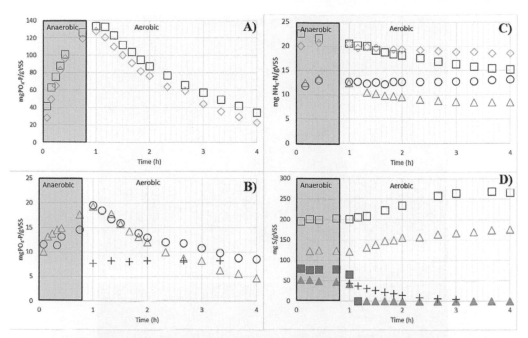

Figure 4.2.- Profiles of the compounds of interest observed in the batch tests performed with sludge from the parent EBPR reactor illustrating the concentrations of: phosphorus (A and B); ammonia (C); and sulphur compounds (D; sulphide concentrations with open squares and sulphate with closed squares). The batch tests were conducted with the addition of: sulphide and VFA (square); only VFA (diamond); and, only sulphide (triangle). The circle markers show the profiles of the control test performed without VFA and without sulphide, and the cross markers display the profiles of the blank test conducted without biomass.

Chapter 4
Long-term effects of sulphide on the enhanced biological removal of phosphorus:
The symbiotic role of *Thiothrix caldifontis*

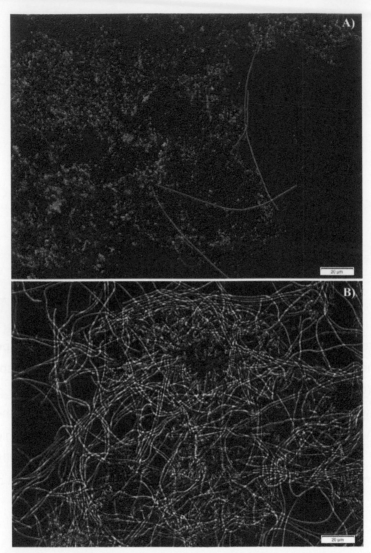

Figure 4.3.- FISH images performed with EBPR sludge from the parent bioreactor after long-term exposure to (A) 10 mgS/L and (B) 100 mgS/L displaying: all living organism in green (DAPI); GAO in blue (GB and DF215,618,988,1020 FISH probes); in red *Candidatus* Accumulibacter phosphatis (FISH probes PAO 462, 651, 846); and, in yellow *Thiothrix* (FISH probe G123T).

Chapter 4
Long-term effects of sulphide on the enhanced biological removal of phosphorus:
The symbiotic role of *Thiothrix caldifontis*

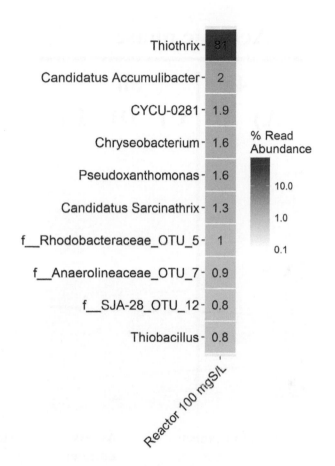

Figure 4.4.- 16 sRNA Amplicon sequencing indicating the read abundance of different organisms present in the EBPR sludge after long-term exposure to 100 mgS/L.

Chapter 4
Long-term effects of sulphide on the enhanced biological removal of phosphorus:
The symbiotic role of *Thiothrix caldifontis*

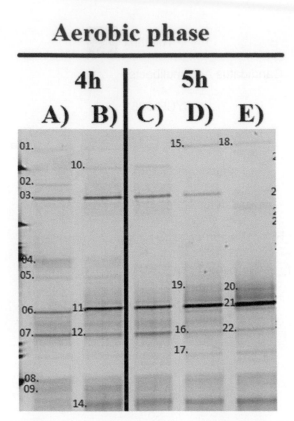

Band number	Related organism	Access number	similarity
3	Uncultured bacterium	DQ413120.1	99 %
6	Uncultured *Thiobacillus*	FJ439098.1	99 %
7,12,16	Uncultured *Thermomonas*	GQ891782.1	99 %
11,21	*Thiothrix caldifontis*	KF926094.1	99 %

Figure 4.5.- DGGE analyses performed with EBPR sludge from the parent bioreactor after long-term exposure with an aerobic phase of 4 h to (A) 10 mgS/L and (B) 20 mgS/L, and after long-term exposure with an aerobic phase of 5 h to (C) 20 mgS/L (D) 50 mgS/L and (E) 100 mgS/L.

Chapter 4
Long-term effects of sulphide on the enhanced biological removal of phosphorus:
The symbiotic role of *Thiothrix caldifontis*

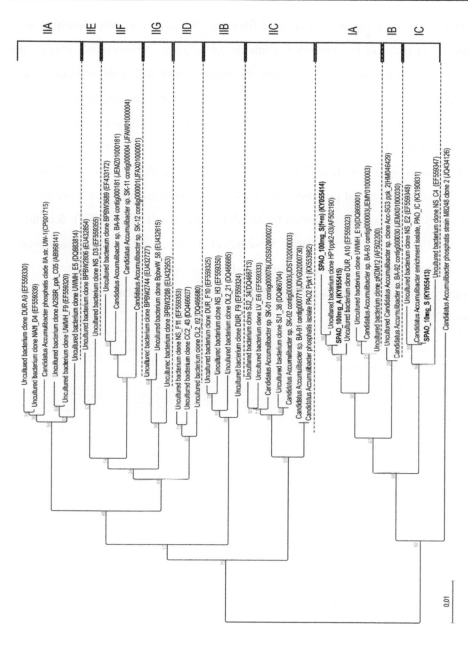

Figure 4.6.- Phylogenetic tree of *Candidatus* Accumulibacter phosphatis based on the ppk gene. Samples SPAO_10_mg and SPAO_100_mg, were taken when the system was in pseudo steady state at 10 and 100 mgS/L fed, respectively.

Chapter 4
Long-term effects of sulphide on the enhanced biological removal of phosphorus:
The symbiotic role of *Thiothrix caldifontis*

The increase in the relative abundance of filamentous bacteria, mostly identified as *T. caldifontis*, led to an increase in suspended solids in the effluent. Nevertheless, despite the deterioration of the settleability of the sludge, the system was still able to achieve full phosphorus removal (Figure 4.1B). Further sludge characterization indicated the existence of different intracellular polymers in the *Thiothrix* organisms: PHA inclusions at the end of the anaerobic stage (Figures 4.7A and 4.7C), while at the end of the aerobic phase Poly-S and Poly-P were detected (Figures 4.7B and 4.7D, respectively). The ability of *T. caldifontis* to take up carbon anaerobically (and possibly store it as PHA) independently of the presence of intracellular Poly-S granules was confirmed by MAR-FISH analysis (Figure 4.8). Similar to the metabolism of PAO (Comeau et al., 1986; Smolders et al.1994, 1995), further microscopic investigations showed that, to some extent, the Poly-P granules present in the *Thiothrix* cells decreased slightly at the end of the anaerobic phase and increased at the end of the aerobic phase (Annex 4.C).

Chapter 4
Long-term effects of sulphide on the enhanced biological removal of phosphorus:
The symbiotic role of *Thiothrix caldifontis*

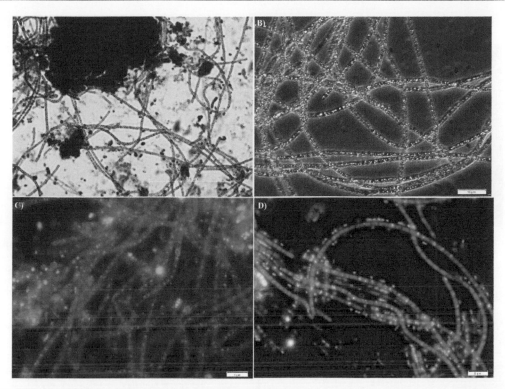

Figure 4.7.- *Thiothrix* images of the EBPR sludge after long-term exposure to 100 mgS/L stained with: (A) Nile blue showing in red PHA inclusions; (B) Contrast image showing in bright white the elemental sulphur inclusions; (C) BODIPY displaying in bright green PHA inclusions; and (D) DAPI staining showing in yellow the intracellular Poly-P inclusions.

Chapter 4
Long-term effects of sulphide on the enhanced biological removal of phosphorus:
The symbiotic role of *Thiothrix caldifontis*

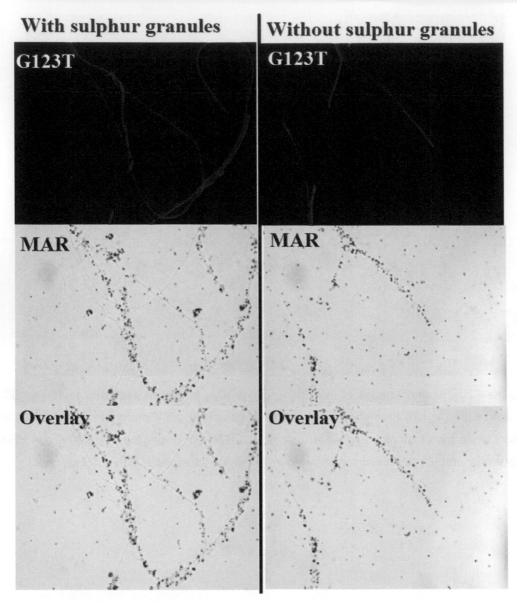

Figure 4.8.- MAR-FISH images carried out with radioactive carbon performed with biomass after long-term exposure to 100 mgS/L with and without the existence of sulphur granules.

Chapter 4
Long-term effects of sulphide on the enhanced biological removal of phosphorus:
The symbiotic role of *Thiothrix caldifontis*

4.6. Discussion

Selection and enrichment of *T. caldifontis*

In this study, the relative abundance of *T. caldifontis* increased as the sulphide concentrations did. As observed by Rubio-Rincon et al. (2016), who reported that 86 mgS/L caused 42% inhibition of the anaerobic carbon uptake of PAO, likely the presence of sulphide inhibited the activity of *Ca.* Accumulibacter while favouring that of *T. caldifontis* which is a sulphide oxidizing organism (Chernousova et al., 2009). Furthermore, *T. caldifontis* was able to store VFA as PHA under anaerobic conditions (Figure 4.7 and 4.8), a process previously proposed for similar cultures but only under aerobic conditions (Schwedt et al., 2012). Thus, when *T. caldifontis* reached the aerobic stage, they rapidly stored the sulphide present in the form of Poly-S and, likely together with the anaerobically stored PHA, they used Poly-S as energy source to support their aerobic metabolic pathways, including biomass synthesis. This is supported by the ammonium consumption observed in the aerobic phase (associated to biomass growth since nitrification did not take place due to the addition of Allyl-N-thiourea) which, interestingly, also increased with the increase in sulphide concentrations. In the same way, an increase the growth yield was observed at higher sulphide concentrations (from 0.18 to 0.39 mg VSS/mg COD_{VFA}, at 10 and 100 mgS/L, respectively), thus higher ammonia consumption.. The anaerobic PHA storage, aerobic Poly-S storage, and further oxidation to sulphate and the aerobic ammonia consumption suggest that *T. caldifontis* grew aerobically following a mixotrophic metabolism as proposed by Chernousova et al. (2009). The ability to grow mixotrophically is well distributed among sulphur oxidizing bacteria, and it is believed to provide an advantage over specialized autotrophic and heterotrophic organisms under alternating growth conditions (Kuenen et al., 1982). Overall, these observations can contribute to explain the proliferation of *T. caldifontis* in the EBPR system initially dominated by *Ca.* Accumulibacter.

Chapter 4
Long-term effects of sulphide on the enhanced biological removal of phosphorus:
The symbiotic role of *Thiothrix caldifontis*

The competition between mixotrophic sulphur oxidizing bacteria might be determined by their ability to oxidize and store intracellular compounds, which should be reflected in the sulphur oxidizing organism that ultimately dominates the system. The faster sulphide oxidation rate in the batch experiments conducted with biomass compared to the control test and a slower sulphate production than sulphide oxidation indicate that sulphide was biologically oxidized and presumably stored as Poly-S. The sulphide oxidation rate observed at 100 mg S/L was 333 mg H_2S-S/gVSS.h (389 mg H_2S-S/L.h), which is considerably higher than the rate observed in *Beggiatoa* cultures by Berg et al. (2014) (3.8 and 18.2 mg H_2S-S/L.h. at 4 and 16 mM sulphide fed). This can explain the dominance of organisms from the *Thiothrix* sp. genus, which is in line with the studies of Nielsen et al. (2000) who observed these organisms can be the dominant filamentous bacteria present in activated sludge plants (with a relative abundance of up to 5-10%).

T. caldifontis contribution to the biological removal of nutrients

In the aerobic stage, complete phosphorus removal was observed at 100 mg S/L when *T. caldifontis* was the dominant bacteria in the system. On the contrary, at 20 mgS/L up to 5 mg PO_4-P/L was observed in the effluent, when *Ca.* Accumulibacter was the dominant bacteria in the system. Rubio-Rincon et al. (2016) reported that sulphide inhibited more severely the aerobic metabolism than the anaerobic metabolism of *Ca.* Accumulibacter. Thus, in this study higher sulphide concentrations should have deteriorated the aerobic phosphorus uptake, which was not the case. It seems that an increase in the abundance of *T. caldifontis* (from 4% to 65% at 10 and 100 mg S/L, respectively) and the rapid oxidization of sulphide observed in the first 10 min of the aerobic stage decreased the sulphide inhibiting effects on the aerobic metabolism of *Ca.* Accumulibacter, contributing to a stable EBPR.

Chapter 4
Long-term effects of sulphide on the enhanced biological removal of phosphorus:
The symbiotic role of *Thiothrix caldifontis*

In addition to the previous observations, *T. caldifontis* might have also directly contributed to the EBPR process by storing phosphorus as Poly-P. According to Wentzel et al. (1989), the maximum storage capacity of *Ca.* Accumulibacter lies around 0.38 mgP/mgVSS. Hence, based on the bio-volume quantification (at 100 mg S/L) of *Ca.* Accumulibacter performed through FISH image analysis of 33%, at an SRT of 4.6 d, and the biomass concentration of the reactor of 1.081 gVSS/L, then around 64.5 mgVSS of *Ca.* Accumulibacter would be removed in each cycle. Thus, considering the maximum Poly-P storage capacity of *Ca.* Accumulibacter and discarding any potential chemical precipitation (since calcium concentrations and Mg/P and K/P ratios were rather typical to those reported for EBPR systems), it can be estimated that *Ca.* Accumulibacter contributed up to a maximum net removal of 24.4 mgPO$_4$-P/L (Annex 4.D). This is insufficient to explain the full net removal of phosphorus observed in the system of 37.5 mg PO$_4$-P/L, suggesting that another organism likely contributed to the biological removal of phosphorus. Thus, possibly *T. caldifontis* (which was the most dominant organism in the EBPR system at 100 mgS/L) stored additional phosphorus beyond their biomass growth requirements, reaching up to 0.10 mg P/mg VSS. The DAPI image analysis for Poly-P detection supports these observations showing certain phosphorus inclusions inside the filamentous bacteria (displayed in yellow or white color) (Figure 4.7, Annex 4.C).

This ability to aerobically store phosphorus could provide to *T. caldifontis* the energetic mechanism to store PHA in the subsequent anaerobic phase and afterwards grow mixotrophically (Kuenen et al., 1982). The mixotrophic growth of *T caldifontis* resulted in a higher ammonia consumption (7.4 and 1.5mg NH$_4$-N/VSS, when sulphide was fed or nor, respectively; figure 4.2C). This indicates that *T. caldifontis* could not solely remove phosphorus and sulphur, but also ammonia due to is mixotrophic growth.

Chapter 4
Long-term effects of sulphide on the enhanced biological removal of phosphorus:
The symbiotic role of *Thiothrix caldifontis*

The ability of sulphur oxidizer organisms to perform the biological removal of phosphorus has been reported in literature. Brock et al. (2011) observed that a *Beggiatoa* culture was capable to intracellularly store phosphorus as Poly-P, using the energy generated by the oxidation of sulphide. However, contrary to this study, Brock et al. (2011) did not observe any PHA accumulation. Hence, *T. caldifontis* may rely on the oxidation of sulphide and PHA to generate energy for the aerobic removal of phosphorus.

Potential anaerobic conversions of *T. caldifontis*

Based on the measurements presented in Table 4.2, assuming that the *Ca.* Accumulibacter population present in the system was capable to store up to 36% of the COD fed to the reactor (as function of the P-released-to-COD consumption ratio)(Annex 4.D), considering all COD as acetate, and the stoichiometry for clade IC reported by Welles et al. (2015), it is possible to estimate the anaerobic and aerobic conversions of *T. caldifontis* (Table 4.3 and 4.4, respectively). Therefore, *Thiothrix* could have stored up to 64% of the COD anaerobically fed. However, only 47% of the COD fed could be tracked down and identified as an intracellular carbon storage compound (PHB/VFA of 0.12 C-mol/C-mol and PHV/VFA of 0.29 C-mol/C-mol). Interestingly, the ratio PHV/PHB was 2.4 C-mol/C-mol, which is close to the theoretical ratio calculated for the reductive branch of the TCA cycle for GAO (2.5 C-mol/C-mol PHV/PHB; Zeng et al., 2003). Potentially the carbon source could also be stored as another PHA polymer, such as PH_2MV. However, even if it was stored as PH_2MV, only 69% of the COD fed assumed to be stored by *T. caldifontis* could be tracked (assuming a PH_2MV/VFA ratio of 0.17 C-mol/C-mol as reported by Zeng et al., 2003). This might suggest that either part of the carbon was oxidized to carbon dioxide or it was stored as another carbon polymer not detected with the analytical methods applied in this study. This can be supported by the studies of Schulz et al. (2005) who observed that even though *T. namibiensis* was capable of storing acetate under anaerobic conditions, while PHA formation was not observed.

Chapter 4
Long-term effects of sulphide on the enhanced biological removal of phosphorus:
The symbiotic role of *Thiothrix caldifontis*

The transport and storage of acetate requires energy (ATP) and reducing equivalents (NADH) (Smolders et al., 1994b). The anaerobic P-release/VFA-uptake ratio estimated for *Thiothrix* sp. was 0.64 P-mol/C-mol. Thus, the ATP provided by the hydrolysis of Poly-P can be assumed to be higher than that required for the reduction and storage of acetate as PHA at pH of 7.6 (Smolders et al., 1994b). On the other hand, the reducing equivalents should be partially provided by glycogen, since the glycogen consumption to PHA formation ratio was only 0.14 C-mol/C-mol. This is clearly insufficient to provide all the reducing equivalents necessary for PHA formation (Smolders et al., 1994b). The source of NADH for reduction of PHA to PHB/PHV is as yet unclear.

Table 4.3.- Anaerobic conversions estimated for the Thiothrix culture in the cycle conducted with 100 mgS/L.

	Ratio	Unit	Measured	Estimated for Accumulibacter[a]	Estimated for *Thiothrix*[b]
Anaerobic Conversions	PHV/PHB	C-mol/C-mol	0.39	0.07	2.42
	PHV/VFA	C-mol/C-mol	0.21	0.09	0.29
	PHB/VFA	C-mol/C-mol	0.54	1.27	0.12
	PH$_2$MV/VFA	C-mol/C-mol	N.M.	N.M.	0.17[c]
	GLY/VFA	C-mol/C-mol	0.20	0.29	0.14
	P/VFA	P-mol/C-mol	0.64	0.64	0.64
Anaerobic net Conversion	PHV/L	C-mmol/L	1.34	0.20	1.14
	PHB/L	C-mmol/L	3.39	2.92	0.47
	PH$_2$MV/L	C-mmol/L	N.M	N.M.	0.68[c]
	P/L	P-mmol/L	4.00	1.47	2.53
	GLY/L	C-mmol/L	1.25	0.67	0.58
	VFA/L	C-mmol/L	6.25	2.3	3.95

[a] Values obtained based on Welles et al. (2015).
[b] Values calculated according to the bio-volume of *Thiothrix sp.* and assuming that acetate was the main carbon source.
[c] Calculated based on the stoichiometric conversions of the reductive branch of the TCA cycle proposed by Zeng et al. (2003)
N.M. Not measured.

Chapter 4
Long-term effects of sulphide on the enhanced biological removal of phosphorus:
The symbiotic role of *Thiothrix caldifontis*

Table 4.4.- Aerobic net conversions and energy balance estimated for the *Thiothrix* culture in the cycle test fed with 100 mg S/L.

	Ratio	Units	Measured PHA[a]	Measured PHA+ Poly-S	Accu mulib acter	*Thiothrix sp Caldifontis* PHA[a]	PHA[b]	PHA+ Poly-S
Net conversions	PHV	C-mol/L	1.34	1.34	0.2	1.14	1.14	1.14
	PHB	C-mol/L	3.39	3.39	2.92	0.47	2.8	2.8
	PO$_4$-P	P-mol/L	4	4	1.47	2.5	2.5	2.5
	Glycogen	C-mol/L	1.25	1.25	0.67	0.58	0.58	0.58
	Biomass Growth	C-mol/L	3	3	0.99	1.99	1.99	1.99
	Sulphide oxidation	S-mol/L	N.A	2.43	N.A.	N.A	N.A	2.43
Energy balance of the aerobic metabolism	**Source/Use**				**NADH balance**			
	PHA		-[0.74][c]	-[0.74][c]	2.18	-[1.69][c]	1.44	1.44
	Biomass growth		1.84	1.84	0.61	1.22	1.22	1.22
	Phosphate transport		-0.44	-0.44	-0.16	-0.27	-0.27	-0.27
	Glycogen		1.25	1.25	0.67	0.58	0.58	0.58
	Sulphide oxidation		N.A	7.29	N.A	N.A	N.A.	7.29
	Balance		**2.69**	**9.94**	**3.3**	**1.53**	**2.97**	**10.26**
	Source/Use				**ATP balance**			
	Oxidative phosphorylation		4.98	18.38	6.10	2.83	5.49	18.98
	PHA		N.A	N.A	0.48	N.A	0.32	0.32
	Biomass growth		-4.50	-4.50	-1.48	-2.98	-2.98	-2.98
	Phosphate transport		-4.00	-4.00	-1.47	-2.50	-2.50	-2.50
	Glycogen		-1.04	-1.04	-0.82	-0.48	-0.48	-0.48
	Balance		**-[4.56][d]**	**8.84**	**2.81**	**-[3.13][d]**	**-[0.15][d]**	**13.34**

[a] Estimations based on the measured PHA concentrations.
[b] Estimations assuming a ratio of 1 C-mol PHA stored per 1 C-mol of VFA consumed.
[c] Insufficient carbon conditions for biomass growth and glycogen formation.
[d] Energy deficient conditions.

Chapter 4
Long-term effects of sulphide on the enhanced biological removal of phosphorus:
The symbiotic role of *Thiothrix caldifontis*

Under aerobic conditions with 50 and 100 mgS/L, in the first 10 min when sulphide was oxidized, it was not possible to observe any phosphorus uptake (Annex 4.A). Thus, it is suggested that *T. caldifontis* stored sulphide as Poly-S, before storing phosphorus as Poly-P. Once Poly-S was stored, both PHA and Poly-S were possibly used to generate energy (Smolders et al., 1994b; Brock et al., 2011) to cover the different metabolic activities of the bacteria. Table 4.4 shows the net aerobic conversions and energy balance under aerobic conditions calculated based on bio-volume quantification (33% *Ca.* Accumulibacter; 65% *T. caldifontis*), considering the stoichiometry for *Ca.* Accumulibacter IC reported by Welles et al. (2015), and assuming that there is not accumulation over time neither of PHA nor glycogen (Annex 4.D3). While *Ca.* Accumulibacter could obtain their carbon and energy requirements covered from the estimated PHA storage pools, *T. caldifontis* had severe limitations if PHA was their only carbon and energy source. For instance, carbon would be missing for biomass synthesis and/or glycogen formation. Either the required carbon might be provided in the form of an intracellular carbon polymer not measured in this study or *T. caldifontis* could use an alternative inorganic carbon source like CO_2. Moreover, even if a ratio of 1 C-mol PHA stored per C-mol of VFA consumed is assumed to try to explain the metabolism of *Thiothrix,* the ATP produced by the oxidation of PHA will result insufficient to cover their different metabolic activities. Alternatively, the energy required could be provided by the oxidation of the intracellularly stored Poly-S into sulphate. The carbon, reducing power and energy balances estimated for the *Thiothrix* culture (Table 4.4) strongly suggest that this mechanism took likely place. This is fully supported by the potential mixotrophic growth that these organisms can perform (Chernousova et al., 2009).

The preference to carry out certain metabolic processes using either PHA or Poly-S can be regulated by the presence or absence of other intracellular polymers (e.g. Poly-P). In the batch test fed with VFA both of them (with and without sulphide) exhibit a similar phosphorus profile, whereas only the one fed with sulphide had a higher ammonia consumption (2.0 and 5.4 mg NH_4-N/gVSS, respectively). This indicates, that as Poly-P needs to get restored

Chapter 4
Long-term effects of sulphide on the enhanced biological removal of phosphorus:
The symbiotic role of *Thiothrix caldifontis*

(due to its anaerobic release), growth is performed at last with the energy that remains from the PHA and/or Poly-S oxidation. On the other hand, in the batch test fed with sulphide, twice as much phosphorus was removed and a higher ammonia consumption was observed than in the batch conducted without sulphide (Figure 4.2). In this case, Poly-P was not hydrolysed as much as compared when VFA was added (anaerobic phosphorus release), thus as less phosphorus has to be replenish more energy can be used for growth.

4.7. Conclusions

Thiothrix caldifontis, a mixotrophic sulphide oxidizing organism, contributed to the stability and performance of an EBPR system fed with sulphide up to 100 mgS/L that achieved full P-removal. Also, *T. caldifontis* directly contributed to the removal of phosphorus via aerobic P-uptake, accumulating up to 100 mgP/gVSS. *T. caldifontis* was able to perform (i) the anaerobic storage of VFA as PHA, (ii) aerobic sulphide storage as Poly-S, (iii) mixotrophic growth using Poly-S and PHA, (iv) and aerobic phosphorus uptake. This research suggests that *T. caldifontis* potentially can be used for the biological removal (and possible later recovery) of phosphate and sulphur.

5

Cooperation between Competibacter sp. and Accumulibacter in denitrification and phosphate removal processes

Chapter 5
Cooperation between Competibacter sp. And Accumulibacter in denitrification and
phosphate removal processes

5.1. Highlights

- A PAO I culture did not perform a significant anoxic P-uptake using nitrate.

- A PAO I-GAO culture showed a higher anoxic P-removal activity on nitrate.

- The PAO I and PAO I-GAO cultures were able to perform anoxic P-removal using nitrite.

- GAO contributed to the reduction of nitrate to nitrite thereby supporting the anoxic activity of PAO

Adapted from

Rubio-Rincón F.J.[a,*], Lopez-Vazquez C.M.[a], Welles L.[a], van Loosdrecht M.C.M.[b], Brdjanovic D.[a,b] (submitted) Cooperation between Competibacter sp. and Accumulibacter in denitrification and phosphate removal processes.

Chapter 5
Cooperation between Competibacter sp. And Accumulibacter in denitrification and
phosphate removal processes

5.2. Abstract

The use of nitrate/nitrite as electron acceptor during the biological removal of phosphate helps to reduce the carbon requirements and sludge production of wastewater treatment plants. Although simultaneous P-removal and nitrate reduction has been observed in laboratory studies as well as full-scale plants, several studies indicate that both PAO I and PAO II are only capable of using nitrite as external electron acceptor. This suggests that other bacterial populations are responsible for the partial denitrification step from nitrate to nitrite. The denitrification capacities of two different cultures, a highly enriched PAO I and a PAO I-GAO culture were assessed through batch activity tests conducted before and after acclimatization to nitrate. Three different electron acceptors (oxygen, nitrate, and nitrite) in combination with two carbon sources (acetate and propionate) were used to assess the biomass activity. Negligible phosphate removal was observed in the highly enriched PAO I culture when nitrate was the electron acceptor, on the opposite the PAO I-GAO culture showed a higher anoxic P-uptake activity. Both systems exhibited good anoxic P-uptake capacity using nitrite as electron acceptor, without requiring any acclimatization or adaptation step. These findings suggest that other organisms capable to reduce nitrate into nitrite but usually believe to be undesirable (e.g. GAO), may play a crucial role in denitrifying EBPR systems. Moreover, the simultaneous denitrification and P-removal process using nitrite as electron acceptor may be a more stable and sustainable process, able to be established by enhancing a nitritation step process in wastewater treatment plants.

Keywords: EBPR; DPAO; PAO I; GAO; DGAO; denitrification.

Chapter 5
Cooperation between Competibacter sp. And Accumulibacter in denitrification and
phosphate removal processes

5.3. Introduction

Phosphate is a key nutrient to remove from wastewater streams to avoid eutrophication
of water bodies (Yeoman et al., 1988). One of the ways to remove phosphate is biologically,
through the use of polyphosphate accumulating organisms (PAO) in wastewater treatment
plants (WWTP). *Candidatus* Accumulibacter phosphatis are one of the main PAO performing
the biological removal of phosphate in WWTP (Hesselmann et al., 1999). Under anaerobic
conditions, PAO are able to store volatile fatty acids (VFA) as polyhydroxyalkanoates (PHA),
at the expense of the polyphosphate hydrolysis and glycogen conversion. Under aerobic or
anoxic conditions, PAO oxidize stored PHA to generate energy which is used to replenish
polyphosphate and glycogen, to grow, and for maintenance purposes (Smolders et al., 1994a,
1994b; Kuba et al., 1996b). The ability of PAO to store VFA as PHA during anaerobic
conditions and use it during oxic/anoxic conditions, give them an advantage to proliferate over
ordinary heterotrophic organisms (OHO) (Henze et al., 2008).

The ability to store VFA as PHA under anaerobic conditions is not restricted to PAO.
Other organisms rely solely on the consumption of glycogen (and therefore do not contribute
to the biological removal of phosphate) as source of energy for the storage of VFA (*Candidatus*
Competibacter phosphatis and *Defluvicoccus* hereafter referred to as glycogen accumulating
organism ; GAO), they have been observed to proliferate and coexist with PAO in EBPR
systems (Cech et al., 1993). As GAO do not directly contribute to P-removal, their existence
is generally associated to the failure of P-removal performance in enhanced biological
phosphorus removal (EBPR) systems (Oehmen et al., 2007). Nevertheless, the ecological role
of GAO in EBPR communities may be more diverse than just being a competitor of PAO.

Most of the WWTP that perform EBPR also remove nitrogen through a
nitrification/denitrification processes. Early studies suggested that PAO could perform the
anoxic uptake of phosphate using nitrate or nitrite as electron acceptors, minimising the
requirements of carbon and sludge production (Kerrn-Jespersen et al., 1993; Kuba et al., 1993).

Such anoxic phosphate uptake was also observed in the anoxic stages of full-scale WWTPs, confirming the existence and role of denitrifying poly-phosphate accumulating organisms (DPAO) (Kuba et al., 1997a, 1997b; Kim et al., 2013). Later studies indicated that the ability to use nitrate or nitrite in denitrification was related to the PAO clade (Ahn et al., 2001a, 2002; Zeng et al., 2003b).

Using both the 16SrRNA gene and the poly-phosphate kinase gene (ppk1) as a genetic marker, past research indicated that Accumulibacter is organized in two main clades: *Candidatus* Accumulibacter phosphatis clade I (PAO I) and *Candidatus* Accumulibacter phosphatis clade II (PAO II). Both clades composed of several distinct sub-clades (McMahon et al., 2002; Seviour et al., 2003a; He et al., 2007; Peterson et al., 2008). Interestingly, a metagenomic analysis indicated that the genome of a PAO culture (presumably PAO II) lacked the respiratory nitrate reductase enzyme, but contained the mechanisms to denitrify from nitrite onwards (García Martín et al., 2006).

Later studies carried out by Carvalho et al. (2007), Flowers et al. (2009) and Oehmen et al. (2010a,b) suggested that the PAOI clade could use nitrate and nitrite as an electron acceptor, but the PAO II clade could solely use nitrite. Lanham et al. (2011) assessed the ability of PAO I to use nitrate and nitrite as electron acceptor using a highly enriched culture (with about 90% PAO I, whereas GAO were not detected) and concluded that PAO I was able to use both nitrate and nitrite as electron acceptor. On the contrary, other studies demonstrated that highly enriched PAO I cultures cultivated under anaerobic-anoxic-oxic (A2O) conditions were not able to perform an anoxic P-uptake activity (Kim et al., 2013; Saad et al., 2016), whereas others, where PAO II was dominant, were able to achieve anoxic P-removal (Guerrero et al., 2012; Wang et al., 2014).

These contradicting findings suggest that other factors rather than the occurrence of a specific PAO clade affect the denitrification capacity of EBPR systems. In this regard, different studies on mix PAO-GAO cultures suggest that GAO rather than PAO use nitrate as electron acceptor (Kong et al., 2006; Lemaire et al., 2006). Still, the actual PAO and GAO fractions

Chapter 5
Cooperation between Competibacter sp. And Accumulibacter in denitrification and
phosphate removal processes

present in those systems (of only 2 - 6 % *Candidatus* Competibacter) and limited knowledge regarding the identity of the dominant PAO clades present in those studies do not allow to reach conclusive observations. In line with these observations, Ribera-Guardia et al. (2016) suggested that PAO has a preference for nitrite over nitrate as electron acceptor. Nevertheless, up to 30 % of the biomass present in that DPAO reactor was not characterized, and the relative abundance of PAO I compared to all bacteria was around 26 %.

Thus, it is still unclear whether PAO I is able to directly use nitrate for the uptake of phosphate or if PAO I rather uses the nitrite generated by side populations, such as GAO. This research aims to understand the role of GAO in denitrifying EBPR systems. For this purpose, the ability to use nitrate or nitrite as electron acceptor of an enriched PAO I and a PAO I-GAO culture were assessed and compared.

5.4. Materials and Methods
Enrichment of the PAO I and PAO I-GAO mix cultures

Two EBPR systems were enriched in two double jacketed reactors with a working volume of 2.5 L. Both reactors were inoculated with 500 mL of activated sludge from Nieuwe Waterweg WWTP (Hoek van Holland, The Netherlands). The reactors were automatically controlled as sequencing batch reactors (SBR) in cycles of 6 h, consisting of 5 min feeding, 2 h 10 min anaerobic phase, 2 h 15 min aerobic phase, 1 h settling time and 30 min for effluent withdrawal. During the effluent phase, half of the working volume was removed providing an HRT of 12 h. At the end of the aerobic phase, 78 mL of mixed liquor were wasted in order to control the SBR at a solids retention time (SRT) of 8 d. The pH was controlled at 7.6 ± 0.1 in reactor 1 and at 7.0 ± 0.1 in reactor 2 through the addition of 0.4M HCl and NaOH. The temperature was controlled externally with a LAUDA system at $20\pm1°C$, and the dissolved oxygen (DO) concentration at 20 % of the saturation level through the addition of compressed air and nitrogen gas. The DO and pH were continuously monitored online, ortho-phosphate (PO_4-P), total suspended solids (TSS) and volatile suspended solids (VSS) were measured twice per week. When no significant changes in these parameters were observed for at least 3 SRT, it was assumed that the system was under pseudo steady-state conditions.

Synthetic media

The media was concentrated 10 times and separated in two bottles of 10 L. The first bottle contained the carbon source, in the case of reactor 1 it comprised 63.75 g of NaOAc•3H$_2$O and 6.675 mL of propionic acid while for reactor 2 it contained 85.44 g of NaOAc•H$_2$O. The carbon source solutions were fed to both systems to reach 396 mg COD/L in the influent of each reactor at the start of the cycle. The second bottle contained 10.7 g NH$_4$Cl, 11.13 g NaH$_2$PO$_4$•H$_2$O, 9 g MgSO$_4$•7H$_2$O, 1.4 g CaCl$_2$•2H$_2$O, 3.6 g KCl, 0.1 g yeast extract, 0.2 g N-allylthiourea (ATU) and 30 mL of trace element solution prepared according to Smolders et al. (1994). After being fed and diluted inside the reactor, the initial phosphate and nitrogen concentrations were 25 mg PO$_4$-P/L and 36 mg NH$_4$-N/L, respectively.

Biomass acclimatization to anoxic conditions

To assess the denitrification activity of the biomass cultures enriched in both systems, short-term activity tests were conducted in the parent reactors and in a batch set-up. Assuming that the nitrite/nitrate reductase enzymes may only be expressed and active after a period of exposure to anoxic conditions (Flowers et al., 2009), the biomass in the reactors was acclimatized to such conditions prior to the execution of the tests. In order to acclimatize the biomass and at the same time ensure the complete uptake of phosphate, the cycle of the reactor was modified from anaerobic-aerobic to anaerobic-anoxic-aerobic conditions. The anaerobic phase had a length of 45 min, the anoxic phase of 120 min, the aerobic phase of 120 min, settling time of 45 min and effluent removal of 30 min. To ensure anoxic conditions, nitrate was fed as a pulse for 1 min at a flowrate of 12 mL/min from a stock solution that contained 2 g NO$_3$-N/L (reaching 10 mg NO$_3$-N/L in the reactor). The acclimatization of the reactors was carried out for 8 cycles prior to the conduction of each batch activity test.

Chapter 5
Cooperation between Competibacter sp. And Accumulibacter in denitrification and
phosphate removal processes

Batch activity tests

Once the reactors reached a pseudo steady-state, batch tests were conducted before
and after the anoxic acclimatization of the biomass. Depending on the test, either before or
after the acclimatization, 200 mL of TSS were taken after the aerobic phase, diluted up to 400
mL and introduced into two 500 mL double-jacketed reactors. After each sludge transfer, the
sludge waste in the parent reactors was stopped for a certain number of cycles to compensate
for the biomass withdrawal. In order to identify the role of GAO in denitrifying EBPR, either
propionate (which benefit the carbon uptake by PAO) or acetate was used as carbon source
(Pijuan et al., 2004). Both carbon sources were tested in combination with three electron
acceptors (oxygen, nitrate, and nitrite). In each test, the organic concentration was kept at 396
mg COD/L. In the corresponding tests, the concentrations of DO, NO_3-N and NO_2-N were 8.6
mg O_2/L, 5 mg NO_3-N /L and 5 mg NO_2-N/L, respectively. Oxygen was continuously supplied
by sparging compressed air into the reactors, while nitrate and nitrite through the addition of 1
mL of concentrate solutions of either 2 g NO_3-N/L or 2 g NO_2-N/L, respectively. Nitrate and
nitrite were constantly measured using Merck Millipore nitrate and nitrite detection strips,
respectively (Amsterdam, The Netherlands). The pH and temperature were controlled at 7.0 ±
0.1 and 20 ± 1°C. The sludge was constantly stirred at 300 rpm using magnetic bars and stirring
plates. The TSS, VSS, and ash content were measured at the start of the test, end of anaerobic
phase and end of the aerobic/anoxic phase. Acetate and propionate concentrations were
measured during the anaerobic phase. The nitrate, nitrite, and ammonia concentrations were
measured during the anoxic and aerobic phases, and ortho-phosphate was measured throughout
the experiment.

In order to verify the potential role of GAO in the denitrification of EBPR systems, an
extra experiment was developed aiming to enhance the activity of either PAO or GAO. Thus,
the pH during the anaerobic phase was controlled either at pH 6.0 to favour carbon uptake of
GAO or at pH 8.0 to enhance the uptake of carbon by PAO (Filipe et al., 2001). This
experiment was carried out using acetate as carbon source and the PAO I- GAO culture, which

Chapter 5
Cooperation between Competibacter sp. And Accumulibacter in denitrification and
phosphate removal processes

contained similar fractions of PAO I and GAO.

Long-term exposure of PAO I to anoxic conditions (NO_3^-)

In order to enhance the development of a denitrifying EBPR culture that could use nitrate as electron acceptor, the reactor 1 was operated under an anaerobic-anoxic-aerobic (A^2O) configuration for 4 SRT. The 6 h cycle was the same way like during the acclimatization of biomass to the presence of nitrate, but the nitrate concentration was gradually increased up to 20 mg NO_3-N/L. Ortho-phosphate, ammonia, nitrate, nitrite and DO were measured at different time intervals during one cycle.

In order to assess the contribution of endogenous respiration and cell decay to denitrification, 250 mg VSS 1 month old sludge extracted from the same reactor were added to a cycle in the anoxic phase. Moreover, the biomass was hydrolysed for 3 h at pH 2.0 and re-adjusted to pH 7.0 prior to addition to the reactor.

Analyses

Ortho-phosphate and nitrite were analysed according to methods 4500-P-C and 4500-NO_2-B, respectively, as described in APHA (2005). Nitrate and ammonia were measured according to ISO 7890/1 (1986) and NEN 6472 (1983), respectively. Acetate and propionate were measured using a Varian 430-GC Gas Chromatograph (GC) equipped with a split injector (200°C), a WCOT Fused Silica column (105°C) and coupled to a FID detector (300°C). Helium gas was used as carrier gas and 50 µL of butyric acid as internal standard. TSS, and VSS were measured in triplicate, as described in the analytical technique method (APHA et al., 2005).

Chapter 5
Cooperation between Competibacter sp. And Accumulibacter in denitrification and
phosphate removal processes

Fluorescence *in situ* Hybridization (FISH)

To estimate the microbial populations distribution in the reactors, FISH analyses were performed according to Amman (1995). PAO were targeted with the PAOMIX probe (mixture of probes PAO 462, PAO 651 and PAO 846) (Crocetti et al., 2000). The presence of PAO clade I and clade II was assessed with probes Acc-1-444 and Acc-2-444 (Flowers et al., 2009). *Competibacter phosphatis* was identified with the GB probe according to Kong et al. (2002). *Defluvicoccus* clusters 1 and 2 were identified with the TFO-DF215, TFO-DF618, DF988 and DF1020 probes (Wong et al., 2004; Meyer et al., 2006).Vectashield with DAPI was used to amplify the fluorescence, avoid the fading and stain all living organisms (Halkjær et al., 2009).

FISH quantification of each probe was performed by image analysis of 25 random pictures taken with an Olympus BX5i microscope and analysed with the software Cell Dimensions 1.5 (Hamburg, Germany). The relative abundance of bacteria was estimated based on the percentage of surface area positive stained by the corresponding probes with regard to the total area covered with DAPI (Flowers et al., 2009). The standard error of the mean was calculated as described by Oehmen et al. (2010b).

Stoichiometry and Kinetics

The ratio P/VFA was calculated based on the observed net phosphate released at the end of the anaerobic phase per total organic carbon consumed. The rates of interest were:

q_{NO_3} .- Nitrate uptake rate, in mg NO_3-N/gVSS.h

q_{NO_2}.- Nitrite uptake rate, in mg NO_2-N/gVSS.h

q_{PO_4,NO_3} .- Anoxic phosphate uptake rate in the presence of nitrate,

in mg PO_4-P/gVSS.h

q_{PO_4,NO_2} .- Anoxic phosphate uptake rate in the presence of nitrite,

in mg PO_4-P/gVSS.h

$q_{PO_4,Ox}$.- Aerobic phosphate uptake rate, in mg PO_4-P/gVSS.h

Chapter 5
Cooperation between Competibacter sp. And Accumulibacter in denitrification and
phosphate removal processes

All rates were calculated by linear regression based on the profiles observed as described in Smolders et al. (1995). The oxygen uptake rate (OUR) was measured in a separate biological oxygen monitoring (BOM) unit equipped with a WTW OXi 340i unit connected to the software Multilab as described in Lopez-Vazquez et al. (2008).

5.5. Results
Biomass characterization in the EBPR systems

Two EBPR systems (hereafter EBPR1 and EBPR2 for the PAOI and PAO I-GAO cultures, respectively) were operated for more than 150 days. They reached a pseudo steady-state before the activity tests were carried out. Both systems showed complete P-removal and consumed all carbon source within the first 15 min of the anaerobic phase. The anaerobic P-release/VFA-uptake ratio in EBPR1 was significantly higher than in EBPR2 (0.65 and 0.45 P-mmol/C-mmol, respectively), indicating that EBPR1 had a higher PAO fraction. Under aerobic conditions, EBPR1 took up phosphate at a faster rate than EBPR2 (47 mgPO$_4$-P/gVSS.h versus 23 mgPO$_4$-P/gVSS.h, respectively; Annex 5.A).The general steady-state behaviour was in line with previous studies in our laboratory (Saad et al., 2016).

In order to assess the dominant microbial communities involved in each system, FISH analyses (Figure 5.1) showed that the sludge in EBPR1 contained 97 ± 4% Candidatus Accumulibacter *phosphatis* from which more than 99% belonged to Candidatus Accumulibacter *phosphatis* Clade I. The sludge in EBPR2 was composed of a mixed culture of Candidatus Accumulibacter *phosphatis* (47 ± 3%) and Candidatus Competibacter *phosphatis* (47 ± 5%). The fraction of Candidatus Accumulibacter *phosphatis* in EBPR2 consisted mainly of *C.* Accumulibacter *phosphatis* clade I (> 94 %). *Defluvicoccus* was not detected in either reactor, while in contrast to EBPR2 Candidatus Competibacter *phosphatis* was not detected in EBPR1.

113

Chapter 5
Cooperation between Competibacter sp. And Accumulibacter in denitrification and
phosphate removal processes

Batch activity tests before acclimatization to nitrate

Three control tests executed before the acclimatization of the biomass to the presence of nitrate were performed with sludge from the PAO I-GAO culture (EBPR2). The addition of three different electron acceptors (oxygen, nitrate, and nitrite) was assessed in each batch test. In the three tests, all the carbon source (acetate) was consumed in the anaerobic phase, resulting in P-release concentrations of up to 108 ± 12, mgP/L. In the experiment conducted with oxygen, a complete removal of phosphate was observed with a maximum OUR of 47 mgO$_2$/gVSS.h and a phosphate uptake rate of 29.8 mgPO$_4$-P/gVSS.h. During the experiment with nitrite as electron acceptor, the system was able to take up 14.3 mgPO$_4$-P/gVSS.h and 4.8 mg NO$_2$-N/gVSS.h. In the experiment conducted with nitrate as electron acceptor, neither phosphate nor nitrate removal was observed.

Chapter 5
Cooperation between Competibacter sp. And Accumulibacter in denitrification and
phosphate removal processes

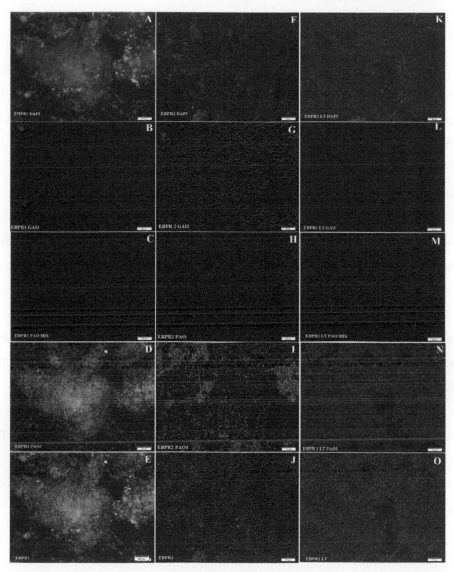

Figure 5.1.- Microbial identification analyses performed by fluorescence *in situ* hybridization
(FISH) in the biomass cultures cultivated in reactor EBPR 1 (A to E), EBPR 2 (F to J), and
after the long-term exposure to nitrate in EBPR 1 (K to O). The green color indicates DAPI
staining, blue GAO mix, red PAO mix and yellow PAO I.

Chapter 5
Cooperation between Competibacter sp. And Accumulibacter in denitrification and
phosphate removal processes

Batch activity tests after acclimatization to nitrate

Effect of acetate as carbon source

The anoxic and aerobic P-uptake capacity, after adaptation to nitrate presence in both cultures was assessed using three electron acceptors: oxygen, nitrate, and nitrite. During anaerobic conditions, acetate was fully consumed in all tests. However, as expected, the anaerobic P-release was considerably higher in the batch tests performed with the PAO I culture (EBPR1) than in the tests performed with the PAO I-GAO culture (EBPR2): 151 ± 11 mgPO$_4$-P/ L and 115 ± 1 mgPO$_4$-P/L, respectively.

Table 5.1 shows the aerobic/anoxic phosphate uptake and nitrate/nitrite reduction rates observed in the PAO I and PAO I-GAO cultures. A similar P-uptake profile was observed under aerobic conditions in all tests. The PAO I culture had a 49 % faster phosphate uptake rate than the PAO I-GAO culture (Table 5.1). The maximum OUR was slightly higher in the mixed PAO I-GAO culture than in the PAO I enriched culture (51 mgO$_2$/gVSS.h and 44 mgO$_2$/gVSS.h, respectively). The net ammonia consumption (assumed to be directly associated to grow) was 4 mgNH$_4$-N/L in the PAO I culture and 5 mgNH$_4$-N/L in the PAO I-GAO culture.

When nitrate was used as electron acceptor, it was not possible to observe any considerable anoxic P-uptake in the PAO I culture (Table 5.1). On the other hand, the PAO I-GAO culture removed 8.7 mgPO$_4$-P/gVSS.h together with 3.2 mgNO$_3$-N/gVSS.h. Nitrite never accumulated in any test as a potential (intermediate) product of the denitrification process (Annex 5.B).

Chapter 5
Cooperation between Competibacter sp. And Accumulibacter in denitrification and
phosphate removal processes

Table 5.1.- Maximum specific phosphate uptake rates of the PAO I and mixed PAO I-GAO cultures with three different electron acceptors (oxygen, nitrate or nitrite) and two different carbon sources (acetate, and propionate).

Culture	C-source	$q_{PO_4,Ox}$ mgP/gVSS.h	q_{PO_4,NO_3} mgP/gVSS.h	q_{PO_4,NO_2} mgP/gVSS.h	q_{NO_3} mgN/gVSS.h	q_{NO_2} mgN/gVSS.h
PAO I	Acetate	37.8	0.6	9.6	1.8	11.4
	Propionate	32.5	3.4	7.7	2.6	11.1
PAO I-GAO	Acetate	25.4	8.7	7.7	3.2	6.5
	Propionate	13.6	7.7	11.5	5.7	7.5
PAO I[a]	Acetate	29.7	4.7	N.A	3.9	N.A

a Observed rate after 4 SRT under anaerobic-anoxic-oxic(A2O) operation conditions.

In order to assess the anoxic P-uptake activities on nitrite of both cultures another set of tests was carried out under anaerobic-anoxic conditions. The PAO I culture displayed a faster P-uptake rate of 9.6 mgPO$_4$-P/gVSS.h compared to 7.7 mgPO$_4$-P-gVSS.h observed in the PAO I-GAO culture. Like previously, the observed nitrite uptake rate of the PAO I culture was faster than the observed nitrite uptake rate of the PAO I-GAO culture (11.4 mgNO$_2$-N/gVSS.h and 6.5 mgNO$_2$-N/gVSS.h, respectively).

Effect of propionate as carbon source

In order to assess the effects of propionate as carbon source (assumed to favour PAO over GAO; Oehmen et al., 2005), the same sets of tests described above were repeated but adding propionate instead of acetate. In all of the tests, propionate was fully consumed under anaerobic conditions. During the aerobic period, both cultures were able to remove all phosphate completely but at different rates. The PAO I culture was capable to take up phosphate and oxygen at a rate of 32.5 mgPO$_4$-P/gVSS.h and 58 mgO$_2$/gVSS.h, respectively. The P-uptake rate and maximum OUR of the PAO I-GAO culture was around half of those obtained with the PAO I culture: 13.6 mgPO$_4$-P/gVSS.h and 27 mgO$_2$/gVSS.h, correspondingly.

Chapter 5
Cooperation between Competibacter sp. And Accumulibacter in denitrification and
phosphate removal processes

In contrast, when nitrate was used as an electron acceptor the P-removal activity of the
PAO I-GAO culture was substantially higher than that of the PAO I culture. The anoxic
phosphate uptake and nitrate reduction rate observed in the PAO I culture were 45 % of those
observed in the PAO I-GAO culture (Table 5.1). On the contrary, when nitrite was added as
electron acceptor the PAO I culture had a faster nitrite reduction rate than the PAO I-GAO
culture (11.1 and 7.5 $mgNO_2$-N/gVSS.h). However, the anoxic P-uptake over nitrite of the
PAO I-GAO culture was substantially higher than the anoxic P-uptake over nitrite of the PAO
I culture (11.5 and 7.7 $mgPO_4$-P /gVSS.h, respectively).

pH effects

Because the PAOI/GAO culture showed a higher denitrifying activity on nitrate than
the PAO I culture, two extra tests were carried out to further assess the role of GAO in
denitrification with the PAO I-GAO culture. The pH during the anaerobic phase of each test
was adjusted to 6.0 or 8.0 which, according to Filipe et al. (2001), should benefit the acetate
uptake of either PAO (at pH 8.0) or GAO (at pH 6.0). During the corresponding following
anoxic phases (using nitrate as electron acceptor) the pH was maintained at 7.0 in both tests.
The anaerobic P/VFA ratio was smaller at pH 6.0 than at pH 8.0 (0.47 and 0.76 P-mmol/C-
mmol, respectively) in line with previous observation on the pH effect on PAO cultures
(Smolders et al., 1994b; Filipe et al., 2001).

In the anoxic test that followed the anaerobic incubation at pH 6.0, the maximum
anoxic P-uptake rate observed was 6.5 $mgPO_4$-P/gVSS.h (Figure 5.2), whereas after the
anaerobic incubation conducted at pH 8.0, the anoxic P-uptake rate showed two different
trends. In the anaerobic incubation conducted at pH 8.0, an initial anoxic P-uptake rate of 3.7
$mgPO_4$-P/gVSS.h was observed in the first hour and afterwards a slower anoxic P-uptake rate
of only 1.2 $mgPO_4$-P/gVSS.h (Figure 5.2). The same trend was observed in its nitrate uptake
rate, decreasing by 43 % from 3.6 to 2.1 $mgNO_3$-N/gVSS.h, as the pH increase from 6.0 to 8.0
in the PAO I-GAO culture.

118

Chapter 5
Cooperation between Competibacter sp. And Accumulibacter in denitrification and
phosphate removal processes

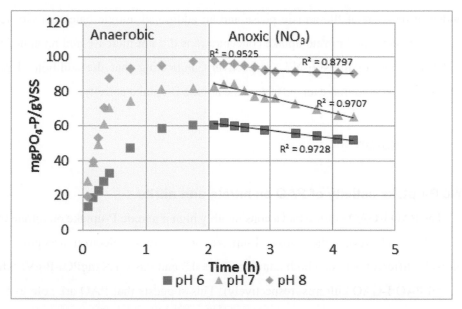

Figure 5.2.- Anoxic phosphate uptake profiles at pH 7.0 observed in the mix PAO I-GAO culture (EBPR2) after different anaerobic stages performed at pH 6.0, 7.0 and 8.0 using acetate as carbon source.

Long-term anaerobic-anoxic-aerobic (A₂O) performance

To assess if the long-term exposure of the system to nitrate could enhance the anoxic P-removal performance, the SBR containing the PAO I culture was operated in an anaerobic-anoxic-aerobic (A₂O) cycles for 4 SRT. A microbial characterization was performed to study if the long-term exposure could favour the growth of other microorganisms, in addition to PAO I. Nevertheless, a relevant change in the dominant microbial populations was not observed (Figure 1). The fraction of PAO I remained above 90% and that of GAO below 5%. 21 mgPO₄-P/L were anoxically removed with a P-uptake rate of 4.7 mgPO₄-P/gVSS.h. The nitrate uptake rate increased to 3.9 mgNO₃-N/gVSS.h, removing up to 17 mgNO₃-N/L. To explain if such nitrate consumption could be result of biomass decay, 250 mg VSS of 1 month old biomass

Chapter 5
Cooperation between Competibacter sp. And Accumulibacter in denitrification and
phosphate removal processes

was added at the start of the anoxic phase and an additional anoxic batch activity test was
conducted. The net P and nitrate uptake concentrations did not increase and remained around
20 mgPO$_4$-P/L and 15 mgNO$_3$-N/L, respectively, indicating that the addition of partially
digested biomass did not have any effect on the anoxic P-uptake activity.

5.6. Discussion

Anoxic P-uptake activity of PAO on nitrate and nitrite

The PAO I-GAO culture had a considerably higher anoxic P-uptake on nitrate than the
PAO I culture. However, the anoxic P-uptake on nitrite as electron acceptor was not
considerably different between both cultures (8.7 ± 0.3 and 9.6 ± 1.8 mgPO$_4$-P/gVSS.h in the
PAO I and PAO I-GAO cultures, respectively). This suggests that PAO are able to denitrify
over nitrite, but that they may rely on other species to denitrify over nitrate. Moreover, Tayà
et al. (2013) proposed to use nitrite as a selective measure for PAO I in the so called PAO-
GAO competition since GAO were unable to denitrify over nitrite, supporting the observations
that PAO use nitrite rather than nitrate for the anoxic uptake of phosphate. This is in line with
the study of McIlroy et al. (2014), which observed that not all GAO have the nitrite reductase
enzyme (nir).

Based on lab- and full-scale observations, Kerrn-Jespersen et al. (1993) and Kuba et
al. (1997b) described the existence of denitrifying polyphosphate-accumulation organisms
(DPAO) able to use nitrate as electron acceptor. Later on, using microbial and molecular
techniques, the existence of two PAO clades was proposed: PAO I, able to utilize nitrate as
electron acceptor and therefore capable to performing the anoxic removal of phosphate, and
PAO II, unable to use nitrate as electron acceptor (Flowers et al., 2009; Oehmen et al., 2010b;
Lanham et al., 2011). Lanham et al. (2011), using an enriched PAO I (90 %) culture, concluded
that PAO I is able to use nitrate for P-uptake ($\pm$ 9 mgPO$_4$-P/gVSS.h, observed in a graphic).

Chapter 5
Cooperation between Competibacter sp. And Accumulibacter in denitrification and
phosphate removal processes

Nevertheless, once the maximum anoxic phosphate uptake was reached, they observed a moderate anoxic phosphate release together with a slightly glycogen consumption, whereas the biomass still contained PHA (around 0.5 C-mmol/gVSS). Thus, as PHA and nitrate are still present, it might suggest that the bacteria could not generate energy from the oxidation of PHA with nitrate and, instead, start to generate it from the hydrolysis of poly-phosphate and glycogen consumption. Furthermore, around 10 % of the biomass in their studies, was composed of rod shaped bacteria belonging to alpha- and gamma-proteobacteria (possibly GAO), which might be able to reduce nitrate into nitrite.

The results obtained in this research indicate that the PAO I culture enriched in our system, cannot efficiently use nitrate as electron acceptor and therefore is unable to perform an efficient anoxic P-uptake activity on nitrate (Table 5.1), which is in agreement with previous studies (Guisasola et al., 2009; Saad et al., 2016). Similar to the observations obtained in our research (Figure 5.2), Lanham et al. (2011) reported that a pH increase led to a lower anoxic P-uptake per mol of nitrate (Table 5.2). According to Filipe et al. (2001), a pH above 7.2 benefits the acetate uptake by PAO.

Chapter 5
Cooperation between Competibacter sp. And Accumulibacter in denitrification and
phosphate removal processes

Table 5.2.- Comparative stoichiometric ratios reported in literature from different EBPR
systems performing anoxic P-uptake activities as a function of pH and Accumulibacter
fractions.

Source	pH	Carbon	P/VFA	$molO_2/$ mol P	$molNO_3-N$ /molP	$molNO_2-N$ /molP	% Acc.	% PAOI
This study	7	Acetate	0.54	0.65	0.73-0.88	0.94-1.58	50±3%	47±3%
This study	7	Propionate	0.58	0.54	0.86-0.97	1.11-1.35	50±3%	47±3%
This study	6	Acetate	0.40	N.R.	0.76-1.12	N.R.	50±3%	47±3%
This study	8	Acetate	0.70	N.R.	0.83-5.00	N.R.	50±3%	47±3%
This study	7	Acetate	0.72	0.45	5.16[a]	1.18-3.8	98±3%	97±4%
This study	7	Propionate	0.52	0.54	0.84-1.53[a]	1.19-1.95	98±3%	97±4%
Carvalho et al. (2007)	7.0-8.2	Acetate	0.52		5.00		64±2%	
Carvalho et al. (2007)	7.0-8.2	Acetate	0.16		1.67		37±2%	
Carvalho et al. (2007)	7.0-8.2	Propionate	0.4		0.80		89±2%	
Carvalho et al. (2007)	7.0-8.2	Propionate	0.32		1.22		76±2%	
Kuba et al. (1993)		Acetate	0.58	1.15	1.07			
Lanham et al. (2011)	7.1-7.2	Propionate	0.53		1	1.12	90%	
Lanham et al. (2011)	7.5	Propionate	0.53		1.33	N.R.	90%	
Lanham et al. (2011)	7.9	Propionate	0.53		1.59	N.R.	90%	
Lanham et al. (2011)	8	Propionate	0.53		1.72	1.53	90%	
Flowers et al. (2009)	7.3	Acetate	0.62		1.59		72±11%	93±1%
Flowers et al. (2009)	7.3	Acetate	0.62		3.45		82±11%	39±1%
Vargas et al. (2011)	7.5	Propionate	0.38			2.96±0.34	60±4%	NR
Vargas et al. (2011)	7.5	Acetate	0.55			2.07±0.40	40±7%	NR

Chapter 5
Cooperation between Competibacter sp. And Accumulibacter in denitrification and
phosphate removal processes

Thus, if PAO I were able to efficiently use nitrate, this should have led to a higher anoxic P-uptake whereas the opposite should have occurred at lower pH. The limited anoxic P-uptake activity at pH 8.0 but higher anoxic P-removal performance at pH 6.0 suggests that GAO carry out the denitrification process from nitrate to nitrite, and that PAO I denitrify from nitrite onwards. The two different anoxic P-uptake rates observed at pH 8.0 in this study (Figure 5.2) support this hypothesis as an increase of pH just slows down the acetate uptake by GAO (Filipe et al., 2001). Thus, the PHA stored by GAO at pH 8 might had been used at the start of the anaerobic phase to denitrify and supply nitrite to PAO I for anoxic P-uptake (3.7 mgPO$_4$-P/gVSS.h) but became limiting after 1 h (1.2 mgPO$_4$-P/gVSS.h) once the carbon source of GAO got exhausted. These are strong indications that GAO could have an essential role in the simultaneous denitrification and phosphate removal processes.

Effect of the carbon source on the anoxic P-uptake activity

Similar maximum specific anoxic P-uptake rates were observed in the PAO I and PAO I-GAO culture when either acetate or propionate was added as carbon source and nitrite was present as electron acceptor (8.7 ± 0.3 and 9.6 ± 1.8 mg PO$_4$-P/gVSS.h, respectively; Table 1). This is in line with previous studies performed by Vargas et al. (2011), who obtained similar anoxic P-uptake rates when either propionate or acetate was used as carbon source (12.7 and 14.8 mgPO$_4$-P/gVSS.h, respectively) and nitrite as electron acceptor. The addition of propionate and nitrate to the PAO I culture increased the anoxic P-uptake from 0.6 to 3.4 mgPO$_4$-P/gVSS.h (Table 1). This is in agreement with the observations of Carvalho et al. (2007) who reported a higher anoxic P-uptake rate when propionate instead of acetate was used as carbon source and nitrate as electron acceptor (19.5 and 8.4 mgPO$_4$-P/gVSS.h, respectively). Carvalho et al. (2007) suggested that the replenishment of glycogen played an important role in the stability and performance of the acetate and propionate fed reactors. However, our experiments were executed at short term and performed with similar initial biomass composition and characteristics. Therefore, the initial intracellular content of glycogen could

123

Chapter 5
Cooperation between Competibacter sp. And Accumulibacter in denitrification and
phosphate removal processes

be assumed to be similar, but not the fraction of PHA. Thus, the fraction of PHA (PHV/PHB) might play an important role, and during anoxic conditions PAO could have preferential pathways depending on which storage polymer is more essential to be restored. The use of acetate as carbon source requires more glycogen than when propionate is fed, which might result in a preferential pathway to restore glycogen under anoxic conditions, resulting in a lower phosphate uptake as observed by Carvalho et al. (2007).

Role of flanking communities in denitrification

Compared to this study, Lanham et al. (2011) observed a similar stoichiometry (at pH 7.0) when using propionate as carbon source (Table 5.2). This suggests that both cultures had similar fractions of PAO I (on day 183 in the study performed by Lanham et al., 2011). However, as previously explained, a considerable anoxic P-uptake on nitrate was not observed in our study. On the other hand, Skennerton et al. (2014) reported that the enzymes used for denitrification (nitrate reductase and periplasmic nitrate reductase enzymes) differ among the subclades of PAO I. Thus, the main differences observed between our studies and the ones of Lanham et al. (2011) might be due to either the fractions of flanking communities (side populations) or the subclades of PAO I present in both systems.

In a similar study, Flowers et al. (2009) suggested that PAO I was able to use nitrate as electron acceptor (Table 5.2). However, the uptake rates reported by Flowers et al. (2009) of 1.4 $mgNO_3$-N/gVSS.h and 2 $mgPO_4$-P/gVSS.h can be considered practically negligible (lower than the ones found in this study). Moreover, the flanking communities present in their sludge could account for up to 20 to 30 % of the total microbial populations (based on their reported estimations) and therefore the presence of GAO cannot be discarded. The authors suggest that due to the anaerobic P/VFA ratio of 0.61 P-mol/C-mol it was unlikely that GAO were present. However, this ratio is lower than the one observed in our study (of 0.72 ± 0.05 P-mol/C-mol), making feasible the presence of GAO (or other PHA organisms able to store PHA anaerobically) in their system.

124

Chapter 5
Cooperation between Competibacter sp. And Accumulibacter in denitrification and
phosphate removal processes

Interestingly, Kim et al. (2013) observed a decrease in the *Accumulibacter* fraction from 55 to 29 % and an increase of *Decholoromonas* from 1 to 19 % and Competibacter from 16 to 20% when decreasing the length of the aerobic phase and increasing the anoxic phase. Furthermore, the increase in the dose of nitrate resulted in an increase in the anoxic P-uptake activity, and led to the accumulation of nitrite (and then consumption). These observations are in agreement with the higher anoxic P-uptake activity observed in the mixed PAO I-GAO culture, supporting the hypothesis that GAO contribute to the anoxic P-uptake activity by denitrifying the available nitrate to nitrite for its further utilization by PAO.

The hypothesis that GAO (or other side communities) are essential for the anoxic P-uptake of PAO on nitrate is in agreement with the observations drawn by Garcia Martin et al. (2006). They suggested that the first part of the nitrate respiration might be carried out by flanking communities since the PAO culture of that study lacked the nitrate reductase enzyme but had the rest of the required enzymes to perform the denitrification process from nitrite onwards. Besides GAO, other flanking communities, even ordinary heterotrophs, could satisfy their carbon needs on dead biomass or ex-polymeric substances (Ichihashi et al., 2006a). Fermentative PAO like *Tetraspera* (Kristiansen et al., 2013) and autotrophic organisms able to use other electron donors and acceptors (inorganic carbon, methane or sulphide) (Brock et al., 2012), could also play a role in EBPR systems on the first denitrification step from nitrate to nitrite on the benefit of PAO.

Implications for full-scale systems

The PAO I community enriched in this study could take up phosphate efficiently under anoxic conditions using nitrite as electron acceptor. When nitrification occurs to nitrate, it means that other organisms such as GAO have to perform the partial denitrification step to nitrite in order to sustain an efficient anoxic P-uptake and EBPR process. This could be a reason for the regularly observed instabilities observed in EBPR performances of full scale WWTP. For carbon and energy efficient and stable EBPR processes, it might be beneficial to

Chapter 5
Cooperation between Competibacter sp. And Accumulibacter in denitrification and
phosphate removal processes

integrate EBPR with a partial nitritation process. This approach could reduce the carbon consumption (which could potentially be diverted to biogas production), oxygen supply, and due to the lower anoxic growth yield, contribute to a lower sludge production. Also this might be an alternative for a partial nitritation Anammox process (Mulder et al., 1995), which would rely on full BOD removal in the first treatment stage. With a first stage, removing stably a large fraction but not all BOD, the combination of EBPR over nitrite might be more attractive than only Anammox applications since simultaneous nitrogen and phosphate removal can be achieved as described in recent promising full-scale observations by Yeshi et al. (2016).

5.7. Conclusions

The enriched PAO I culture had a slower uptake of phosphorus and nitrate when compared to the anoxic conversions on nitrite. In the PAO I-GAO culture the differences in the anoxic phosphate uptake rates between nitrate and nitrite as electron acceptors were lower. This suggests that not all PAO I can fully denitrify and that GAO might not only compete with PAO for substrate in the anaerobic period, but also supply electron acceptors (nitrite) in anoxic environments to PAO in a partly competitive and partly syntrophic relationship.

6

Absent anoxic activity of PAO I on nitrate under different long-term operational conditions

6.1. Highlights

- Negligible anoxic phosphorus uptake activity was observed in a highly enriched PAO IC culture.

- With a 15 d SRT, marginal denitrification activity was observed without phosphorus uptake.

- At a lower organic loading, phosphorus was not taken up but released under anoxic conditions.

- A short aerobic SRT of 0.4 d led to the deterioration of the system.

Adapted from

Rubio-Rincón F.J., Welles L, Lopez-Vazquez C.M., Weissbrodt D.G., Abbas B, Geleijnse M., van Loosdrecht M.C.M., Brdjanovic D. (In preparation) Absent anoxic activity of PAO IC on nitrate under different long-term operational conditions

6.2. Abstract

"Candidatus Accumulibacter phosphatis" clade I, a known polyphosphate-accumulating organism (PAO I), has been proposed to be able to perform anoxic P-uptake activity in enhanced biological phosphorus removal (EBPR) systems using nitrate as electron acceptor. However, different studies disagree on their ability to use nitrate. One of the reasons of such discrepancies might be related to the diverse operational conditions applied across the different studies. This study aimed to assess whether and how certain operational conditions could enhance the denitrification capacity of PAO I. Thus, an EBPR culture enriched with PAO I was operated for a long-term period (weeks) under anaerobic–anoxic–aerobic (A$_2$O) conditions. The media composition, solids retention time (SRT), polyphosphate content of the biomass (Poly-P), nitrate addition mode, and minimal aerobic SRT were specifically studied. Despite the different attempts and range of conditions applied, only a marginal rate of anoxic P-uptake was observed, equivalent to a maximum of 13% of the P-uptake rate quantified under aerobic conditions (5 and 39 mg PO$_4$-P/gVSS.h, respectively). Furthermore, an increase in the anoxic SRT at expense of the aerobic SRT resulted in phosphorus present in the effluent and acetate leakage into the anoxic phase. These results suggest that PAO I used in these experiments, is not capable of generating energy (ATP) from nitrate under strict anaerobic-anoxic conditions as efficient as with dissolved oxygen under anaerobic-aerobic ones. The reduction of nitrate to nitrite observed in PAO I enriched cultures might be due to nitrate dissimilation by PAO and/or the denitrification activity of side- communities present in the sludge. The single-lineage genomic information recovered from the metagenome of the EBPR sludge indicated that this organism did not harbour the respiratory nitrate reductase (*nar*) but only the periplasmic nitrate reductase (*nap*) gene. These results suggest that nitrite and not nitrate was used to provide the energy needed for the anoxic dephosphatation observed in the past.

Keywords: *"Candidatus* Accumulibacter", PAO, DPAO, PAO I, Denitrification, Anoxic phosphorus uptake

6.3. Introduction

Enhanced Biological Phosphorus Removal (EBPR) is a worldwide applied process used to remove phosphorus in wastewater treatment plants (WWTP) (Henze et al., 2008). EBPR is carried out by microorganisms broadly known as polyphosphate accumulating organisms (PAOs) that are capable of storing phosphorus beyond their growth requirements as poly-phosphate (Poly-P) (Comeau et al., 1986; Mino et al., 1987). PAOs have different metabolic processes depending on the availability of electron acceptors. Under anaerobic conditions, PAOs store volatile fatty acids (VFA) (e.g. acetate, propionate) as poly-hydroxy-β-alkanoates (PHA) at the expense of poly-P hydrolysis and glycogen degradation. Thereafter, when an electron acceptor is available (e.g. oxygen, nitrate, nitrite), PAOs consume the stored PHA to replenish their Poly-P and glycogen storage pools, for biomass synthesis and maintenance purposes (Comeau et al., 1986; Wentzel et al., 1986; Smolders et al., 1994a, 1994b; Kuba et al., 1996b).

Previous studies have suggested that certain types of PAO could use nitrate or nitrite as electron acceptors for anoxic P-uptake contributing to reduce the carbon requirements and the sludge production of WWTP (Kerrn-Jespersen et al., 1993; Kuba et al., 1993, 1996a, 1997a, Ahn et al., 2001a, 2001b). Kerrn-Jespersen et al. (1993) studied the potential use of nitrate as electron acceptor for the biological removal of phosphorus through several anaerobic-anoxic batch tests. They postulated the existence of two types of PAO: one able to use nitrate and oxygen as electron acceptor (herein identified as denitrifying-PAO or DPAO) and another PAO capable of using only oxygen. Likewise, through the long-term operation of two sequencing batch reactors (SBR) operated under anaerobic-anoxic (A2) and anaerobic-aerobic conditions (A/O), Kuba et al. (1993) observed that DPAO could have a similar anoxic EBPR activity like PAO on oxygen.

Carvalho et al. (2007) observed the dominance of a rod-shape PAO on a propionate-fed reactor (HPr) that exhibited satisfactory anoxic P removal on nitrate and the proliferation of coccus-shape PAO on an acetate-fed (HAc) reactor that showed poor anoxic EBPR activity.

Later on, Flowers et al. (2008) identified the existence of two clades of Candidatus *Accumulibacter phosphatis*: *clade I* and *clade II* based on the presence of the ppk1 and ppk2 genes on two PAO cultures (hereafter referred to as PAO I and PAO II, respectively). Based on the findings of Flowers et al. (2008), Oehmen et al. (2010) assessed the different PAOs observed by Carvalho et al. (2007) and suggested that PAO I was responsible for the anoxic P-uptake activity observed in EBPR systems. Furthermore, Flowers et al. (2009) observed that a PAO I enriched culture (composed of 70 ± 11 % PAO I with regard to DAPI staining) was able to denitrify without requiring any acclimatization step, while a PAO II enriched culture (55±7% PAO II/DAPI) could not. Nevertheless, after the acclimatization of both EBPR systems to the presence of nitrate for 24 h, both reactors could perform a simultaneous denitrification and phosphorus uptake activity. The anoxic P-uptake activity observed in the PAO II enriched reactor was associated to the presence of PAO I (32±0.5 % PAOI/DAPI) (Flowers et al., 2009).

Lanham et al. (2011) enriched a PAO I culture (± 90% abundance) under anaerobic-anoxic-aerobic (A2O) conditions that was capable to take up around 12 mg PO_4-P/gVSS.h. while using nitrate as electron acceptor However, contrary to the merely A/O operating conditions of Kuba et al. (1993), to maintain the aerobic stage seemed to have been a key condition to secure the reactor stability and obtain a highly enriched PAO I culture. In contrast, through the execution of short-term studies, Saad et al. (2016) and Rubio-Rincon et al. (chapter 5) suggest that PAO I (and more specifically, PAO IC) is unable to use nitrate as electron acceptor as efficient as oxygen, and even than nitrite. As such, they suggest that when anoxic P-uptake takes place, PAO I may use the nitrite generated from the reduction of nitrate by side communities, possibly glycogen-accumulating organisms (GAO) or ordinary heterotrophs organisms. Nevertheless, both studies were based on the conduction of short-term (hours) batch activity test with PAO biomass enriched under A/O conditions. On the other hand, recent studies performed by Camejo et al. (2016) suggest that PAO IC can use nitrate as electron acceptor. Nevertheless, even if the main clade was PAO IC (>92%) their culture comprised only 15 to 20% *Candidatus* Accumulibacter phosphatis. Since the fraction of PAO seems to be rather low it cannot be discarded that other microbial communities rather than PAO perform the first denitrification step from nitrate to nitrite.

From a microbial perspective, metagenomic analyses of PAO IA and IC reveal the absence of the respiratory nitrate reductase enzyme (*nar*) required for nitrate respiration (Flowers et al., 2013; Skennerton et al., 2014) and they only have a similar denitrification pathway like PAO IIA and IIF. That pathway only includes the presence of the periplasmic nitrate reductase enzyme (*nap*) and the required enzymes to denitrify from nitrite onwards (García Martín et al., 2006; Flowers et al., 2013; Skennerton et al., 2014). As such and in view of the contradictory findings, the gene regulation mechanisms could be responsible for the different anoxic P-uptake activities of PAO I and EBPR systems reported in literature, as suggested by Skennerton et al. (2014).

The anoxic phosphorus uptake activities observed in past studies could be due to the creation of certain operating conditions that might have triggered or induced an adequate regulation of the denitrifying genes of PAO and the existence and activation of the required denitrifying enzymes. Thus, this study aims to assess the influence of different operating conditions previously applied on a PAO I enriched culture (Rubio-Rincon et al., chapter 5). The EBPR was enriched in PAOI as described by Rubio-Rincon et al., (2016) under A/O conditions, once PAOI was enriched the EBPR was operated for a longer term (weeks) under A2O conditions under the following conditions: (i) initially, the same synthetic media used by Kuba et al. (1993) but with a higher P/COD ratio to benefit PAO over GAO; (ii) a total solids retention time (SRT) similar to that applied by Ahn et al. (2002) assuming that DPAO are slower growing microorganism that aerobic PAO; (iii) thereafter, the influent P/COD ratio was decreased to allow the growth of GAO (Mino et al,. 1998) since they appear to be involved in the reduction of nitrate to nitrite on the benefit of DPAO (Rubio-Rincon et al., chapter 5); (iv) also, the nitrate dosing mode was modified from pulse mode to continuous addition to assess if DPAO could be k- or r-strategist organisms; and, (v) finally, the aerobic SRT was shortened below the minimum reported by Brdjanovic et al. (1998) for A/O EBPR systems with the aim of forcing the washout of aerobic PAO and favour the enrichment of DPAO. This study can contribute to assess whether such conditions can stimulate and enhance the use of nitrate by PAO I and elucidate their potential role in EBPR systems performing satisfactory simultaneous denitrification and phosphorus removal.

6.4. Material and methods

Reactor operation

A culture of "*Ca.* Accumulibacter phosphatis" clade I was enriched in a 3-L double jacket Applikon reactor (Delft, The Netherlands) with a working volume of 2.5 L. A volume of 500 mL of activated sludge from WWTP Nieuwe Waterweg (Hoek van Holland, The Netherlands) was used as inoculum. Prior to the start of this study, the reactor was operated under anaerobic-aerobic conditions for more than a year (experimental period A; Rubio-Rincon et al., chapter 5). Thereafter, the operational conditions were changed to anaerobic-anoxic-aerobic (experimental period B; Table 1). Once the system was operated under anaerobic-anoxic-aerobic conditions in a pseudo steady state (experimental period B), the media composition was change from the one used by Smolders et al., (1994) to the one used by Kuba et al. (1993) (experimental period C). In order to give the opportunity to the assumed slower-growing DPAO to proliferate, the SRT was extended from 8 to 15 d in the experimental period D. Due to the potential role of GAO in the denitrification activities observed in EBPR systems (Rubio-Rincon et al., chapter 5), the feeding P/COD ratio was decreased from 0.06 to 0.03 mg P/mg COD in the experimental period F. In the experimental period G, while keeping the same nitrate concentration ($\pm$ 10 mg NO_3-N/L), the nitrate dosing mode was changed from pulse feeding to continuous feeding during 30 min at 1mL/min flow (as studied by Kuba et al., 1993; Tu et al., 2013) to address if DPAO could be r- or k-strategists. Finally, in order to washout the aerobic PAO on the potential benefit of DPAO, the aerobic SRT was shortened sequentially from 2.2 days to 0.9 days (experimental period G), and further to 0.4 days (experimental period H) which is below the minimum estimated by Brdjanovic et al. (1998) of about 1.25 d.

The hydraulic retention time (HRT) was 12 h. pH was adjusted to 7.6±0.1 with the addition of 0.4 M NaOH and 0.1 M HCl. Temperature was controlled at 20±1°C. In order to create and maintain the anaerobic conditions, nitrogen gas was sparged at the bottom of the reactor during the first 30 min of the anaerobic phase and a water lock was installed at the outlet of the off-gas. Nitrate was fed either as pulse or continuously (according to the experimental period) from a bottle containing a stock solution with 1 g NO_3-N/L. The dissolved oxygen (DO) concentration was controlled at 20% of the saturation level using compressed air and nitrogen gas. Both gases were controlled at 10 L/h. DO and pH levels were monitored continuously. Ortho-phosphate (PO_4-P), total mixed liquor suspended solids (TSS) and mixed liquor volatile suspended solids (VSS) concentrations were measured twice per week. When no significant changes in these parameters were observed for more than 3 SRTs, it was assumed that the system had reached pseudo steady-state conditions.

Table 6.1.- Operational conditions of the different experimental periods.

Period	Days on operation	Operation A$_2$O (anaerobic-anoxic-aerobic)	mg P /L	MLVSS / MLSS ratio	Nitrate concentration mg NO$_3$-N/L (addition mode)	Total SRT (d)	Anoxic SRT (d)	Aerobic SRT (d)
A	N.A.	2h - 0h - 2.5h	25	0.60	N.A.	8	N.A.	3.0
B	0-40	0.6h -2h - 1.8h	25	0.57	20 mg/L (pulse)	8	2.6	2.2
C	40-102	0.6h - 2h -1.8h	25	0.60	14 mg/L (pulse)	8	2.6	2.2
D	102-167	0.6h - 2h -1.8h	25	0.57	13 mg/L (pulse)	15	5.0	4.1
E	167-212	0.6h - 2h -1.8h	25	0.60	11 mg/L (pulse)	8	2.6	2.2
F	212-218	0.6h - 2h -1.8h	15	0.77	11 mg/L (pulse)	8	2.6	2.2
G	218-265	0.6h - 3h -0.8h	15	0.77	30 min at 1 mL/min (continuous)	8	4.0	0.9
H	265-276	0.6h-3.5h-0.3h	15	N.A.	Between 30 and 80 min at 1 mL/min (continuous)	8	4.4	0.4

Nitrate based DPAO batch Activity Tests

In order to assess if the organic load fed per biomass ratio (food-to-microorganisms, F/M ratio) influences the fractions of intracellular compounds and, consequently, the anoxic phosphorus uptake, two batch tests were carried out with half (batch 1D) and twice as much (batch 2D) the F/M ratio applied in the regular operation of the parent reactor. The media composition was the same used in the parent reactor. Each batch test was performed with 200 mL of biomass (collected and transferred during experimental period D, Table 1) in a double jacket reactor with a 400 mL working volume. The cycle of the batch tests was composed of 1h anaerobic and 4h anoxic. In each batch test, nitrate was fed as a pulse reaching a concentration of around 45 mg NO_3-N/L. Nitrogen gas was continuously sparged at the bottom of the reactor at 10L/h in order to maintain the anaerobic conditions. pH was kept at 7.6±0.1 with the automatic addition of 0.4M HCl and 0.4M NaOH.

Media

The media was prepared in two separate bottles of 10 L (carbon and mineral solution), and concentrated 10 times. The media fed contained per litre 400 mg COD (composed by acetate and propionate supplied in a 3:1 ratio), 4 mg Ca^+, 36 mg SO_4^{2-}, 9 mg Mg^{2+}, 1 mg yeast extract, 20 mg N-allylthiourea (ATU) and 300 µL of trace element solution prepared according to Smolders et al. (1994). In addition, the media for (i) the experimental periods A and B contained per litre 36 mg NH_4^+, 25 mg PO_4-P, 19 mg K^+, and 18 mg Na^+; (ii) for periods C, D and E: 83 mg NH_4^+, 25 mg PO_4-P, 50 mg K^+ and 0 mg Na^+; and, (iii) for F, G and H: 83 mg NH_4^+, 15 mg PO_4-P, 38 mg K^+, and 0 mg Na^+.

Analyses

Ortho-phosphate (PO_4-P), nitrite (NO_2-N), TSS, and VSS were analysed as described in APHA (2005). Nitrate (NO_3-N) was measured according to ISO 7890/1 (1986). Acetate (HAc) and propionate (HPr) were measured using a Varian 430-GC Gas Chromatograph (GC) equipped with a split injector (200°C), a WCOT Fused Silica column (105°C) and coupled to a FID detector (300°C). Helium gas was used as carrier gas and 50μL of butyric acid as internal standard.

Microbial characterization

In order to estimate the relative abundance of the microbiology communities along the different experimental phases, Fluorescence *in situ* Hybridization (FISH) analyses were performed as described by Amman (1995). PAOs were stained with the combination of PAO462, PAO651 and PAO846 probes (Crocetti et al., 2000). To differentiate among the different PAO clades, the probes Acc-1-444 (Clade 1A) and Acc-2-444 (Clade 2A, 2C, 2D) were used (Flowers et al., 2009). Glycogen-accumulating organisms (GAOs) were identified with the GB probe according to Kong et al. (2002). Defluvicoccus 1 and 2 were identified with TFO-DF215, TFO-DF618, DF988, and DF1020 probes (Wong et al., 2004; Meyer et al., 2006). Vectashield with DAPI was used to amplify the fluorescence, avoid the fading and stain all living organisms (Halkjær et al., 2009). FISH quantification of each probe was performed by image analysis of 25 random pictures taken with an Olympus BX5i microscope and analysed with the software Cell Dimensions 1.5. The standard error of the mean was calculated as described by Oehmen et al. (2010b).

Analyses of 16S rRNA gene-based amplicon sequencing .

Sequence analysis

The extracted genomic DNA was subsequently used for a two-step PCR reaction targeting the 16Sr-RNA gene of most bacteria and archaea. For this we used the primers, U515F (5' – GTGYCAGCMGCCGCGGTA - 3') and U1071R (5'-GARCTGRCGRCRRCCATGCA- 3') as used by Wang et al. (2009). The first amplification was performed to enrich for 16s-rRNA genes, for this a quantitative PCR essay was setup. The following chemicals were used, 2x iQ™ SYBR® Green Supermix (Bio-rad, CA, USA),

500nM primers each and finally 1-50ng genomic DNA added per well to a final volume of 20µl. The protocol was as followed; a first denaturation of 95°C for 5min and 20 cycles of 95°C for 30 seconds, 50 °C for 40 seconds, 72 °C for 40 seconds and a final extension of 72 °C for 7 minutes During the second step, 454-adapters (Roche) and MID tags at the U515F primer, were added to the products of step one. This protocol was similar only Taq PCR Master Mix (Qiagen Inc, CA, USA) was used and the program was run for 15 cycles, the template, product from step one, was diluted ten times. After the second amplification, twelve PCR products were pooled equimolar and purified over a agarose gel using a GeneJET Gel Extraction Kit (Thermo Fisher Scientific, The Netherlands). The resulting library was send for 454 sequencing and run in 1/8 lane with titanium chemistry by Macrogen Inc. (Seoul, Korea).

Phylogenetic analysis

After analysis the reads library was imported into CLC genomics workbench v7.5.1 (CLC Bio, Aarhus, DK) and (quality, limit=0.05) trimmed to an minimum of 200bp and average of 284bp. After trimming the sample was de-multiplexed resulting in twelve samples with an average of 7800 reads per sample. A build-it SILVA 123.1 SSURef Nr99 taxonomic database was used for BLASTn analysis on the reads under default conditions. The top result was imported into an excel spreadsheet and used to determine taxonomic affiliation and species abundance.

Molecular analysis of PAO clades by PCR on the *ppk1* functional gene

Genomic DNA was extracted using the Ultraclean Microbial DNA extraction kit supplied by MOBIO laboratories Inc. (CA, USA) according to the manufacturers protocol except that the bead-beating was substituted by a combination of 5 minutes heat at 65°C and 5 minutes beat-beating to ensure maximum yields. To check for quality and quantity, the genomic DNA was loaded onto a 1% agarose gel in 1x TAE running buffer. Analysis of the extracted DNA showed a large high molecular weight fraction and well visible DNA yields in comparison to the Smart ladder (Eurogentech Nederland b.v.).

A direct PCR to identify the PAO clade was performed based on the polyphosphate kinase (*ppk1*) functional gene as described by McMahon et al., (2007). The PCR amplicons were sequenced using ACCppk1-254F and ACCppk1-1376R primers. The phylogenetic tree was constructed using the neighbour joining method as described by Saad et al. (2016).

Stoichiometric and Kinetic Parameters of Interest

The phosphorus released to VFA ratio (P/VFA) was calculated based on the observed net phosphorus released at the end of the anaerobic phase per VFA consumed. The phosphorus content in the biomass was calculated as described in Welles et al. (2015). The anaerobic rates of interest were:

$q_{PO_4,AN}^{MAX}$ Maximum anaerobic phosphorus release rate, in mg PO_4-P/gVSS.h.

$m_{PO_4,AN}$ Anaerobic endogenous phosphorus release observed once VFA were taken up, in mg PO_4-P/gVSS.h.

$q_{PO_4,VFA}$ Anaerobic phosphorus release due to VFA uptake, calculated according to:

$$q_{PO_4,VFA} = q_{PO_4,AN}^{MAX} - m_{PO_4,AN}$$ (Equation 6.1).

q_{VFA}^{MAX} Maximum anaerobic VFA uptake rate observed, in mg COD/gVSS.h.

Additionally, OUR profiles were determined based on the dissolved oxygen (DO) consumption over time. In order to measure the DO consumption, during the aerobic stages the EBPR sludge was recirculated from the parent reactor through a separate 10 mL BOM unit for 2-3 min. Once the DO measurements were stable, the sludge recirculation was stopped and the DO concentration profiles were recorded. The DO concentrations were kept above 2 mg O_2/L by periodically re-starting the sludge recirculation as soon as it approached this concentration. This procedure was repeated along the aerobic phases. The biological oxygen monitor BOM unit was equipped with a WTW OXi 340i DO probe (Germany). The anoxic and aerobic rates of interest were:

$q_{NO_3,Ax}$ Nitrate uptake rate, in mg NO_3-N/gVSS.h.

$q_{NO_2,Ax}$ Nitrite uptake rate, in mg NO_2-N/gVSS.h.

$q_{PO_4,Ax}$ Anoxic phosphorus uptake rate, in mg PO_4-P/gVSS.h.

$q_{PO_4,Ox}$ Aerobic phosphorus uptake rate, in mg PO_4-P/gVSS.h.

All rates were calculated by linear regression based on the observed profiles as described in Smolders et al. (1995).

6.5. Results

Operation of the reactor under Anaerobic-Aerobic (A/O) conditions

The reactor was operated for more than a year under Anaerobic-Aerobic conditions showing a pseudo steady-state performance. All VFA were anaerobically removed during the first 15 min, at a rate of 269 mg COD/gVSS.h (q_{VFA}^{MAX}), with a phosphorus release of 199 mg PO_4-P/gVSS.h ($q_{PO_4,AN}^{MAX}$). Once all VFA were taken up, a P-release rate of 2.5 mg PO_4-P/gVSS.h ($m_{PO_4,AN}$) was observed and assumed to correspond to the anaerobic endogenous P-release. Under the presence of oxygen, phosphorus was taken up at a rate of 58 mg PO_4-P/gVSS.h ($q_{PO_4,Ox}$). The observed ratio of phosphorus taken up per total oxygen consumed was 1.63 mg P/mg O_2 (0.42 P-mol/e⁻).

Long-term operation under Anaerobic-Anoxic-Aerobic (A₂O) conditions

As observed in Figure 6.1, during the anaerobic stage, all VFA were taken up and phosphorus was released at maximum anaerobic rates ($q_{PO_4,AN}^{MAX}$) between 164 and 254 mg PO_4-P/gVSS.h. Once the VFA were consumed, it was possible to calculate the phosphorus released due to maintenance activities ($m_{PO_4,AN}$), which was considerable higher in experimental periods B, C, and E (12±0.8 mg PO_4-P/gVSS.h) than in D, F, and G (1.9±1.7 mg PO_4-P/gVSS.h). The P/VFA ratios were rather stable in the experimental periods B, C, D, and E

(0.66±0.06 mg PO$_4$-P/mg COD), and higher than in F and G (0.50±0.06 mg PO$_4$-P/ mg COD).

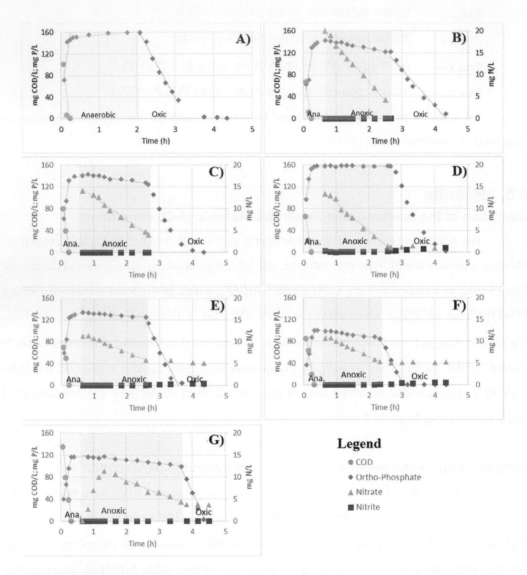

Figure 6.1. Ortho-Phosphate (diamond), VFA as COD (circle), nitrate (square), and nitrite (triangle) concentration profiles observed during the cycles of the different experimental periods assessed.

Table 6.2.- Anaerobic and aerobic rates and stoichiometry calculated along the different experimental periods conducted in this study.

Exp. Period	q_{VFA}^{MAX}	$q_{PO_4,A}^{MAX}$	$m_{PO_4,AN}$	q_{PO_4,NO_3}	q_{PO_4,O_x}	q_{NO_3}	P/NO₃	P/O₂	VSS/TSS[a]
	mgCOD/gVSS.h	mgPO₄-P/gVSS.h				mgNO₃-N/gVSS.h	P-mol/e⁻		g/g
A	269	199	2.5	N.A[d]	58	N.A[d]	N.A[d]	0.42	0.59
B	181	185	13.3	5.0	39	3.9	0.11	0.46	0.62
C	271	240	11.7	4.8	64	3.0	0.13	0.47	0.58
D	150	130	3.8	0.2	50	2.5	0.01[b]	0.53	0.56
E	272	254	12.0	2.9	79	2.0	0.13	0.50	0.61
F	210	164	1.8	5.4	77	2.1	0.24	N.C.[d]	0.77
G	331	193	0.3	3.7	79	1.8	0.22	0.43	0.77
				Activity Test					
1D	75	72	N.O.[d]	-2.7[c]	N.A[d]	0.8	N.A[d]	N.A[d]	0.60
2D	265	235	8.3	0.7	N.A[d]	1.95	0.03	N.A[d]	0.60

a. Calculated at the start of each test.
b. Value below 0.01 P-mol/e⁻.
c. Phosphorus was released instead of taken up.
d. N.O not observed; N.A. not applicable; N.C. not calculated.

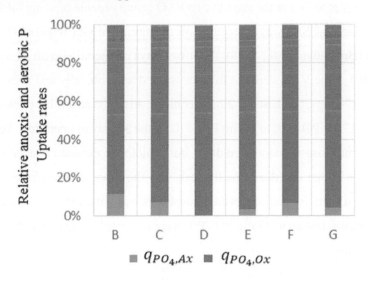

Figure 6.2. . Fractions of the total phosphorus uptake rates related to anoxic (*blue*) and aerobic (*orange*) metabolisms calculated in the different A₂O experiments.

Under anoxic conditions, it was not possible to detect the accumulation of nitrite in none of the experimental periods (Figure 6.1). A slightly higher denitrification activity of 3.9 mg NO_3-N/gVSS.h was observed in the first experimental period (B) after switching to the Anaerobic-Anoxic-Aerobic conditions. However, in the rest of the experimental periods, the denitrification rates did not increased considerably and remained around 2.2±0.47 mg NO_3-N/gVSS.h. Table 6.2 shows the anaerobic-anoxic-aerobic rates and stoichiometry parameters calculated in each experimental period. Compared to the aerobic P-uptake rates (of up to 79 mg PO_4-P/gVSS.h), a relatively low anoxic P uptake rate ($q_{PO_4,Ax}$) was observed in periods B, C, and F (5.0±0.3 mg PO_4-P/gVSS.h), which decreased to 3.7 mg PO_4-P/gVSS.h in period G, or was practically inexistent in period D. As much, the relative anoxic phosphorus uptake rates reached 13% of the aerobic phosphorus uptake rates (Figure 6.2). In all cases, phosphorus was fully removed in the aerobic phases (Figure 6.1).

Minimal aerobic SRT

As an attempt to favour the growth of a PAO group capable of using nitrate over strict aerobic PAO, the aerobic SRT was reduced below the minimum SRT required for aerobic PAO of about 1.25 d (Brdjanovic et al., 1998). After two days of operation with an aerobic SRT of approximately 0.4 d, the VFA were not fully consumed anaerobically and started to leak into the anoxic phase (74 mg COD/L at day 2 of operation). Also, the anaerobic phosphorus release decreased from 117 mg PO_4-P/L to 32 mg PO_4-P/L on the 3th day of operation. To avoid nitrate limitation, the nitrate dose was increased daily, but avoiding to exceed a concentration of more than 3 mg NO_3-N/L that could eventually leak into the next aerobic phase. Despite these measures, no DPAO activity was observed, phosphorus was not removed in neither the anoxic nor the aerobic phase, and up to 27 mg PO_4-P/L were observed at the end of the aerobic phase (day 2 of operation).

Assessment of the effects of the F/M ratio on the anoxic phosphorus uptake activity

In the 1D test (conducted with half the F/M ratio applied to the parent reactor), the acetate uptake rate ($q_{PO_4,VFA}$) and the maximum phosphorus release rate ($q_{PO_4,AN}^{MAX}$) were 75.2 mg COD/gVSS.h and 72.2 mg PO$_4$-P/gVSS.h, respectively. They were considerably slower than the ones observed in the 2D batch test executed with a F/M ratio twice as high the ratio fed to the main reactor (265 mg COD/gVSS.h and 235 mg PO$_4$-P/gVSS.h, respectively). Despite these differences, the observed P/VFA ratios were not considerably different between the two batch tests (0.60 and 0.66 at the 1D and 2D batch tests, respectively).

On the contrary, the anoxic phosphorus uptake and denitrification rates were rather different (Figure 6.3). In the 1D batch test, a nitrate reduction rate of 0.8 mg NO$_3$-N/gVSS.h was observed. But, interestingly, phosphorus was released anoxically at a rate of 2.7 mg PO$_4$-P/gVSS.h instead of being taken up (Figure 6.3A). In contrast, in the 2D batch test a faster nitrate reduction rate of 1.95 mg NO$_3$-N/gVSS.h was observed together with a marginal phosphorus uptake rate of 0.7 mg PO$_4$-P/gVSS.h.

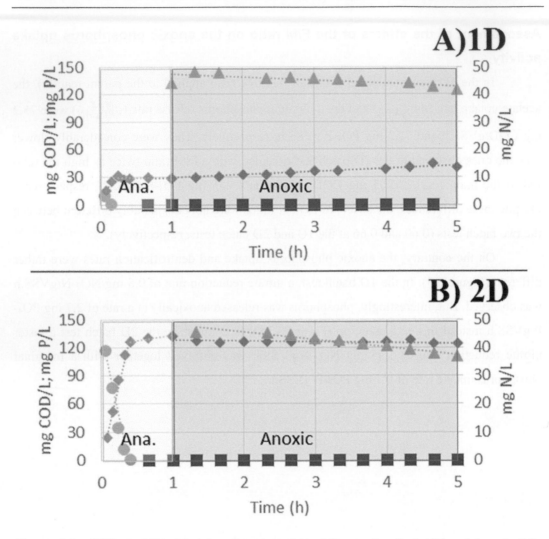

Figure 6.3.- COD as VFA (circle), ortho-phosphate (diamond), nitrate (triangle) and nitrite (square) concentration profiles observed in the activity test with half F/M (A) and twice as high F/M (B) ratio applied in the parent reactor.

Microbial adaptation along the experimental periods

FISH analyses were performed to identify the predominant microorganisms present in the systems and potentially involved for the biological removal of phosphorus over the different experimental periods. Figure 6.4 shows a representative image of the microbial composition at the start and end of this research (experimental periods A and G, respectively). The relative abundance of PAOs related to "*Ca.* Accumulibacter phosphatis" (PAOmix probe set) compared to all organisms (stained with DAPI) decreased from 98%, 95%, 76% to 52% along the experimental periods A, B, D and G, respectively (Annex 6A). Despite these differences, the fraction of the PAO I clade (Acc-1-444 probe) within PAOs did not change along the experimental periods assessed (97±4%; Figure 6.4 and Annex 6A). In none of the experimental periods, considerable fractions of GAO were detected (<5%, "*Ca.* Competibacter phosphatis" and *Defluvicoccus vanus;* figure 6.4 and 6.5).

The 16S rRNA gene-based amplicon sequencing profiles displayed a decrease in the relative abundance of the genus "*Ca.* Accumulibacter" from 53% to 33% in the experimental period A and G, respectively (figure 6.5). Finer-scale characterization of the "*Ca.* Accumulibacter" clades based on PCR and sequencing analyses of the *ppk1* gene showed that the system was composed mainly of members of the clade PAO IC at the start and at the end of the experimental periods (Figure 6.6).

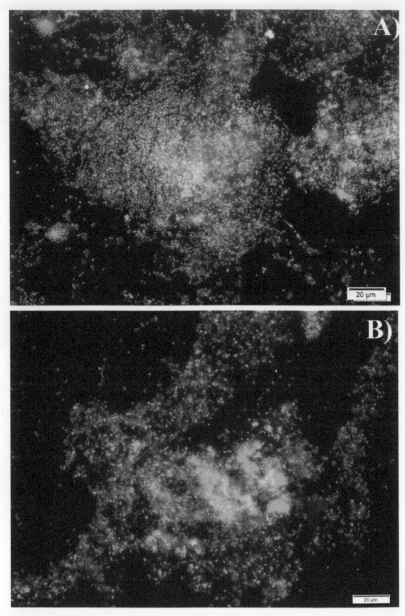

Figure 6.4.- Microbial characterization during experimental period A (A) and experimental period G (B). In green all living organism (DAPI), in red PAO (Cy3), in yellow PAOI (FAM), in blue GAO (Cy5).

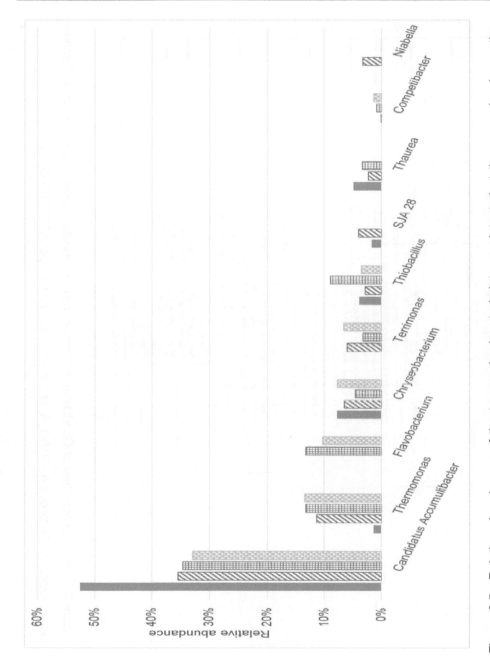

Figure 6.5.- Relative abundances of the ten predominant phylotypes detected at the genus level over the experimental periods A (blue solid), B (orange dash), D (grey squares) and G (yellow dots) by 16S rRNA gene-based amplicon sequencing analysis

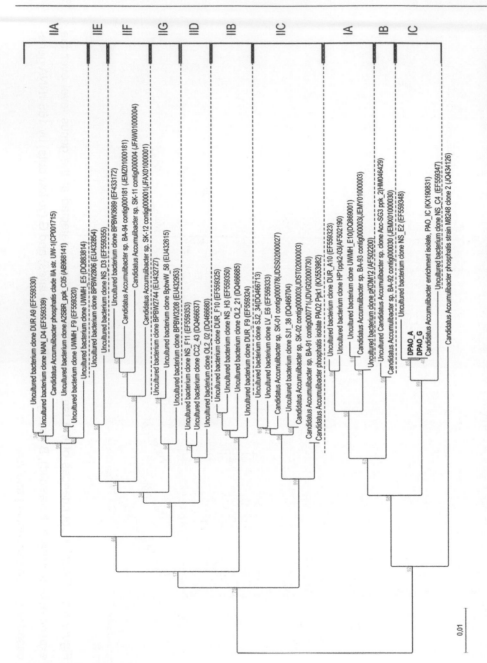

Figure 6.6.- Phylogenetic tree of *Candidatus* Accumulibacter phosphatis based on the ppk gene. Samples during experimental period A (DPAO_A) and during experimental period G (DPAO_G).

6.6. Discussion

Effect of different operating conditions on anoxic phosphorus uptake activity

As observed in Table 6.2 and Figure 6.2, the anoxic phosphorus uptake rates observed in the experimental periods conducted under different operating conditions were considerably lower than the aerobic P-uptake rates. As much, the anoxic phosphorus uptake rate reached 13% of the aerobic phosphorus uptake rate observed (experimental period B; Figure 6.2). This is in agreement with the study of Lanham et al. (2011) who have observed a faster aerobic P-uptake rate than an anoxic one in an A2O reactor, but differs from previous studies where nitrate has been reported as a terminal electron acceptor as efficient as oxygen by PAO for P-uptake (Kuba et al., 1993; Lee et al., 2014). However, Kuba et al. (1993) and Lee et al. (2014) have applied an A_2 configuration in contrast to this study where A_2O configurations were used. This suggests that the aerobic phase applied here at the end of each cycle may have hampered the selection for a PAO population able to use nitrate as efficient as oxygen for anoxic P-uptake.

As the anoxic growth can be up to 70% of the aerobic growth yield (Kuba et al., 1993), the solids retention time was extended from 8 to 15 d in the experimental period D as suggested by Ahn et al. (2002). In contrast to the studies of Kuba et al. (1993) and Ahn et al. (2002), any anoxic P-uptake was never observed. Only a limited anoxic activity of 2.5 mg NO_3-N/gVSS.h was noticed. It may be concluded that either the denitrification activity observed here was not carried out by PAO or this PAO population was not capable to restore its Poly-P storage pools under the applied conditions. Moreover, further activity test with the biomass of experimental period D showed that under a low F/M ratio phosphorus was surprisingly anoxically released at 2.7 mg PO_4-P/gVSS.h. This release of phosphorus strongly suggests that the PAO could not provide enough energy from the putative reduction of nitrate, even though nitrate was present at all time during the activity test (figure 6.3A). Thus, as suggested by Rubio-Rincon et al. (chapter 5) PAO may have used nitrite (potentially generate by side communities) as electron acceptor for the oxidation of PHA. The nitrite produced by side communities would be lower according to a decrease in the organic load per biomass (F/M), resulting in energy limitation by the oxidation of PHA by PAOs and a subsequent anoxic P-release as observed in figure 6.3A.

A side community (*e.g.* GAO) could have been responsible for the low nitrate reduction activity (as suggested by Rubio-Rincon et al., chapter 5). To test this hypothesis, the P/COD ratio in the influent was decreased as an aim to increase the GAO fraction in the system from experimental period F onwards. Clade PAO IC remained the main PAO present ($97\pm4\%$) that decreased from 98% to 52% from experimental periods A to G (Figure 6.4), similar to the study of Carvalho et al. (2007). However, GAO did not proliferate as expected (up to 1% in the experimental period G; figure 6.5). Likely, the relatively high pH of 7.6 limited the growth of GAO (Smolders et al., 1995). At the same time, an increase in the fraction of potentially denitrifying *Thermomonas* and *Chryseobacterium* genera was observed (13% and 8%, respectively in experimental period G; figure 6.5) after the switch from A/O to A$_2$O conditions. Despite that *Thermomonas* and *Chryseobacterium* are capable to reduce both nitrate and nitrite (Mergaert et al., 2003; Kundu et al., 2014), the denitrification activity did not further increase together with their selection (2.1 and 1.8 mg NO$_3$-N/gVSS.h in experimental periods F and G, respectively). The non-PAO-based denitrification activity measured in the system remains puzzling. Additional mRNA-based functional gene expression analyses within the nitrogen cycle would have helped to identify the bacterial populations involved. The role of the side-communities (e.g. *Thermomonas, Flavobacterium, Chryseobacterium, Terrimonas*) that composed the rest of the bacterial populations remains unclear, but under the operating conditions applied in this research they did not significantly contribute to the anoxic P-uptake activity. This suggests that there are other factors or other microorganisms should be present (e.g. GAO was suggested by Rubio-Rincon et al., chapter 5) to enhance the anoxic P-uptake activity over nitrite in EBPR cultures.

Kuba et al. (1993) enriched a DPAO culture under A_2 conditions, dosing nitrate continuously during the anoxic stage. This dosing mode could benefit k-strategists DPAO. Thus, in period G nitrate was dosed continuously during the anoxic phase as described earlier. Certain increase in the anoxic P-uptake per electron available was observed from 0.13 to 0.24 P-mol/e⁻ (between periods F and G, respectively). The anoxic 0.24 P-mol/e⁻ ratio observed in period G was comparable to the one of 0.19 P-mol/e⁻ reported by Kuba et al. (1993). However, the anoxic P-uptake rate was lower than the one reported by Kuba et al. (1993) (5.4 mg PO_4-P/gVSS.h compared to 30-46 mg PO_4-P/gVSS.h). This indicated that the dosing mode did not play a role to enhance the anoxic phosphorus uptake activity of the sludge.

As an attempt to favour the growth of DPAO over strict aerobic PAO, the operational conditions were gradually modified from anaerobic-anoxic-aerobic to anaerobic-anoxic (experimental periods G and H). However, once the aerobic SRT was reduced below the minimal required, acetate leaked into the anoxic phase (after just 2 days of operation). Despite that nitrate was not limiting, likely PHA was not completely oxidized during the anoxic and aerobic phases and PAO could not take up sufficient phosphorus for anaerobic VFA uptake. This is in agreement with Lanham et al. (2011) who concluded that an aerobic phase was essential to maintain a good removal of phosphorus, where most of the phosphorus uptake occurred. Therefore, it is suggested that the PAO IC enriched under the conditions applied during this study cannot solely rely on the use nitrate for energy generation for metabolic processes (e.g. via PHA oxidation). Recently, Camejo et al. (2016) reported the anoxic P-uptake using nitrate (11 ± 1.7 mgPO$_4$-P/gVSS.h) of a PAO IC culture with a relative abundance of 20% in the microbial community. However, that EBPR culture was operated under alternation of anaerobic and microaerophilic conditions (Camejo et al., 2016). Thus, during electron acceptor limitation the competition among ordinary denitrifiers and DPAO, might be affected. Furthermore, the recirculation of nitrate (produced at the end of the microaerophilic conditions) into the anaerobic phase will offer the opportunity of ordinary denitrifiers to growth.

Interestingly the conditions applied by Camejo et al., (2016) (pH 7.0 to 7.5; 0.2 mg O_2/L) might favour the formation of nitrite over nitrate (Jianlong et al., 2004). Tayà et al. 2013 suggested that nitrite could be used to enrich PAOs over GAOs. Nevertheless the DPAOs enriched under such conditions could not use nitrate as electron acceptor (Guisasola et al., 2009) in contrast to the study of Camejo et la., (2016). In our case as neither oxygen nor nitrate was limited under the enrichment conditions, this will provided an important advantage to proliferate for organism capable to use oxygen as electron acceptor due to the growth yield is higher and so the energy obtained from the oxidation of PHA (Thauer et al., 1977; Kuba et al., 1993).

Possible metabolic pathway for nitrate reduction observed during this study

In this study, only a marginal nitrate consumption up to 3.9 mg NO_3-N/gVSS.h was observed (in experimental period B) in the enriched PAO IC culture. The consumption of nitrate can be associated to growth (nitrate assimilation), generation of metabolic energy (nitrate respiration; *nar*) and/or dissipation of excess reducing power (nitrate dissimilation; *nap*) (Moreno-Vivián et al., 1999). According to previous studies (Skennerton et al., 2014) clades of "*Ca.* Accumulibacter phosphatis" contain either the nitrate respiration enzyme (*nar;* PAO IIC) or the nitrate dissimilation enzyme (*nap;* PAO IC, IA, IIA, IIF). As observed, only the nitrate dissimilation enzyme (*nap*) has been identified in the methanogenomes of PAO IA and PAO IC (Flowers et al., 2013; Skennerton et al., 2014). Thus, in view of the low anoxic P-uptake activity, it can be concluded that the *nap* enzyme does not contribute significantly to the anoxic P-activity of PAO IC on nitrate.

As described by Moreno-Vivián et al. (1999) the *nap* enzyme seems not to be direct involved neither for the nitrate assimilation or the anaerobic respiration, instead the nitrite generated can be used for those proposes depending on the organism. The physiological role of the *nap* enzyme on denitrifying organisms seems to be not completely understood (Bedzyk et al., 1999). Cardenas et al. (1982) were not capable to observe an anoxic growth on

Rhodobacter sphaeroides in the dark. On the contrary, Bell et al. (1993) observed that *Thiosphaera pantotropha* (mutant M-6) was capable to overproduced the periplasmic nitrate reductase activity allowing the growth over nitrate, though the energy generated through this process was lower when compared with the nitrate respiration. Thus, it seems that the existence and activation of the *nar* enzyme appears to be essential for a significant anoxic dephosphatation using nitrate. Furthermore, it seems that PAOs can generate more energy from the oxidation of PHA reducing nitrite to nitrogen gas than reducing nitrate to nitrite (431 and 216 Kj/mol PHB, respectively at pH 7 and 30% inefficiency ;Rittmann et al., 2001). Therefore, under no limiting electron acceptor conditions PAOs would prefer to use nitrite over nitrate for the generation of energy.

Recently, Camejo et al. (2016) have suggested that a PAO IC culture could fully denitrify from nitrate onwards, which contrasts with the findings of Saad et al. (2016) and this study where not significant anoxic phosphorus uptake was observed. Camejo et al. (2016) reported that their PAO IC culture contained all the enzymes necessary for a full denitrification based on a culture that contained 15 to 20% "*Ca.* Accumulibacter" (with regard to EUB) from which 92 % were PAO IC (resulting in around 14 to 18% PAO IC with regard to the total bacterial community). If that PAO IC culture contained all the required enzymes for denitrification from nitrate onwards, it is still unclear what operating conditions are required to enhance the presence and activation of the required enzymes.

Moreover, based on fine scale genetic sequences of the *ppk* gene is possible to differentiate among the "*Ca.* Accumulibacter" lineage (McMahon et al., 2007). Despite that these clades seems to have the same encoding for denitrification, they exhibits different anoxic P-uptake capabilities on nitrate (e.g. anoxic activities PAO IC and IIA)(García Martín et al., 2006; Flowers et al., 2009; Oehmen et al., 2010a; Oehmen et al., 2010b; Flowers et al., 2013; Skennerton et al., 2014; Camejo et al. 2016; Ribera-Guardia et al., 2016). In view of these observations, the classification of the denitrifying PAO communities in terms of the *ppk* gene as either PAO I (with full denitrification capacity from nitrate onwards) or for PAO II (from nitrite onwards)(Flowers et al., 2009; Oehmen et al., 2010a) does not seem to be longer valid and supported. Further studies should then focus on the detection of the key denitrifying enzymes (e.g. *nar*) on top of clade differentiation using the *ppk* marker gene.

It should then be elucidated how different operating conditions can enhance the selection for DPAOs, the activation of their genetic signatures, and their catabolic regulation depending on terminal electron acceptors available. The influence of the operation on genetic regulation (Skennerton et al., 2014) and interaction with the different metabolic pathways and energy consumption deserves special attention. Kuba et al. (1993) have observed that the total electrons transfer to nitrate or oxygen per mol of phosphorus taken up were similar (e.g. 0.19 and 0.23 P-mol/e⁻, respectively). On the contrary, in this research the P-mol/e⁻ ratios using nitrate as electron acceptor varied between 0.5% to 51% compared to oxygen (Table 6.2). The PAO metabolism has been simplified to four main processes in the presence of an electron acceptor: Poly-P and glycogen replenishment, growth, and maintenance (Comeau et al., 1986). Thus, if the stoichiometric P-mol/e⁻ ratio is assumed to be constant, the lower observed P-mol/e⁻ ratio ratios could indicate that PAO under limiting electron acceptor conditions (e.g. when using nitrate) prefers to restore its Poly-P pool over others metabolic process. In line with this, Carvalho et al. (2007) observed a decrease on the replenishment of glycogen during the anoxic phase when compared to the aerobic phase. Further research is needed to elucidate if DPAO have preferential metabolic pathways under anoxic conditions and how these can affect, or be affected by the gene regulations.

Possible sources of carbon for denitrification by side communities

If ultimately DPAO are not able to denitrify from nitrate up to relevant and applicable rates, it remains possible that the anoxic P-uptake observed in EBPR systems is carried out partially (from nitrate to nitrite) by side populations. However, such probable heterotrophic denitrifiers should have access to a carbon source. This adds up to former questioning that has been made on factors (*e.g.*, organic substrate availability and substrate storage under anaerobic conditions) underlying the presence of a diversity of potential denitrifiers in the bacterial community of EBPR processes (Weissbrodt et al., 2014). In the past, research on EBPR has reported the leakage of certain amount of total organic carbon (not comprised by VFA) as a function of the HRT (Ichihashi et al., 2006b). The total organic carbon (TOC) present in the anoxic stage could be potentially used by ordinary heterotrophic organisms (OHOs) as carbon source for denitrification proposes. Kuba et al. (1993) have observed around 5 mg TOC/L leaking into an anoxic phase, which have led to up to 3.8 mg NO_3-N/L reduction (if nitrate was reduced to nitrite). Moreover, nitrate (from the previous anoxic zone) intruded in the following anaerobic stage, independently of the experimental periods. Thus, OHOs could grow in the system on the VFA fed and nitrate available contributing to the direct denitrification from nitrate to nitrite and even in the subsequent anoxic phase merely with the endogenous respiration process. The presence of nitrate during anaerobic conditions have nevertheless be reported in some cases to favour the growth of DPAO in EBPR systems (Ahn et al., 2001a).

Another possibility is the potential presence of other microorganisms capable to store PHA under anaerobic conditions and to then oxidize it using nitrate. Tu et al. (2013) proposed that under carbon limiting conditions PAO would outcompete other PHA accumulating organism such as GAO. This could be the case in the activity test performed with half organic load per biomass, where it was observed a denitrification of 0.8 mg NO_3-N/gVSS.h together with a anoxic P-release of 2.7 mg PO_4-P/gVSS.h. On the contrary in the activity test performed with double organic load per biomass a denitrification of 1.95 mg NO_3-N/gVSS.h and anoxic phosphorus uptake of 0.7 mg PO_4-P/gVSS.h was observe. Thus, it is suggested that the observed anoxic P-uptake in this study was carried out over the nitrite generated by other PHA accumulating organism.

Ekama et al. (1999) pointed out that alternating anoxic-aerobic conditions stimulated the presence of filamentous bacteria. The appearance of filamentous bacteria has also been reported in the studies of Kerrn-Jespersen et al. (1993), and formed up to 10% of the biomass in the studies of Lanham et al. (2011). Furthermore, early studies of Kuba et al. (1993) also reported considerable amounts of sludge washed out through the effluent due to settling problems, suggesting the presence of filamentous bacteria. These filamentous bacteria (as the normally observed type 021N in wastewater treatment plants) can reduce nitrate into nitrite (Williams et al., 1985; Nielsen et al., 2000) and have also been found as one of the main side populations present in EBPR cultures (Garcia-Martin et al., 2006). Further characterization of the metabolisms of the side populations is needed to elucidate their potential involvement in the anoxic P-uptake activities of EBPR cultures.

6.7. Conclusions

PAO IC did not perform a considerable anoxic phosphorus uptake, which was up to 13% of the observed aerobic phosphorus uptake. Even more, a decrease in the organic load fed per biomass seems to change the anoxic phosphorus uptake into release, which indicates that PAO IC needed to anoxically generate energy from PolyP hydrolysis. Finally, the system collapsed when the aerobic SRT was reduced below the minimum required, suggesting that PAO IC cannot rely solely in the use of nitrate as electron acceptor. Thus, it seems that PAO IC cannot use nitrate to generate energy as efficient as it would use oxygen. Hence, it is highly possible that the observed anoxic phosphorus uptake rate in wastewater treatment plants is perform over the respiration of nitrite, which could be generated by side communities.

7

Outlook and main conclusions

7.1. General conclusions

The increase of phosphorus in water bodies causes eutrophication which finally results in hypoxia. From the phosphorus generated by human activities, 70% is found in wastewater. Therefore, it is important to remove phosphorus from the treated effluent at wastewater treatment plants. Enhanced biological removal of phosphorus is a worldwide use process to remove phosphorous biologically. The biological phosphorus removal (BPR) is carried out by organisms capable to store phosphorus above their growth requirements. These organisms are known as poly-phosphate accumulating organisms (PAOs).

The ability to store intracellular phosphorus above their growth requirements seems to be a characteristic found among different bacteria families. *Candidatus* Accumulibacter phosphatis; however, seems to be the main PAO active in wastewater treatment plants (Saunders et al., 2003). Hence, it is important to understand the metabolic response of *Candidatus* Accumulibacter phosphatis during the different processes in a wastewater treatment plant.

One factor that inhibits the different microbial processes in a wastewater treatment plant (e.g. BPR) is sulphide. Sulphide is formed by the reduction of sulphate under anaerobic conditions. This research differentiates between the sulphide formed in the sewage (short-term exposure) and the sulphide formed in the WWTP (long-term exposure).

The biological removal of phosphorus is achieved by the recirculation of sludge through anaerobic-anoxic/oxic conditions. The latter could inhibit the different anaerobic processes such as sulphate reduction. In case sulphate reducers can proliferate in biological nutrient removal plants the PAO will continuously be exposed to sulphide. The effects that the electron acceptors have on the sulphate reduction process while using different carbon sources were assessed.

Thereafter, an enriched culture of PAO I was exposed, both short (hours) and long term (months) to different concentrations of sulphide. This, to assess the effect of sulphide in the anaerobic and aerobic stages of PAO I. The inhibition caused by sulphide may vary according to the electron acceptor used by PAO. Therefore, the effect of sulphide on the aerobic or anoxic metabolism of PAO I was studied.

To do so, the denitrification potential of an enriched PAO I culture, which according to the literature uses nitrate for the generation of energy (Oehmen et al., 2010a), was assessed. Since no considerable anoxic dephosphatation was observed, an attempt was done to promote the denitrification enzymatic activity as well as to allow a more suitable denitrification PAO (DPAO) to growth. Hence, the bioreactor enriched with PAO I was operated under A_2O conditions for months, changing the composition of substrate, solids retention time, and the method of nitrate dosing.

The outcomes of this research can increase the insights in the biological removal of phosphorus, as it helps to understand the possible interactions between *Candidatus* Accumulibacter phosphatis and other microbial populations. In section 7.2 the conclusions of the specific research objectives will be addressed. Subsequently, section 7.3 will address the questions that are raised by the outcome of this research and set out ideas for further investigation.

7.2. Specific conclusions

Effects of electron acceptors in the sulphate reduction process

It was observed that regardless of the type of electron acceptor (nitrate, nitrite, oxygen) to which the sulphate reducer bacteria (SRB) was exposed, the sulphate reduction activity resumed until some extent. The recovery of the sulphate reducing activity; however, seems to be higher when lactate was used as electron donor. Thus, sulphate reducers capable of using lactate as electron donor and perhaps other long-chain carbon sources are most likely to grow in biological nutrient removal plants. According to these findings, the recirculation of sludge from the aerobic tank and short anaerobic retention time most likely would inhibit the proliferation of SRB in the WWTP.

The sulphide effects on the physiology of Candidatus Accumulibacter phosphatis Type I

Sulphide affected both anaerobic and aerobic metabolism of C. Accumulibacter phosphatis. However, the aerobic metabolism was more severely affected. It was possible to observe a 50% decrease in the acetate uptake and phosphate release rate at 115 mg TS-S/L. On the other hand, under aerobic conditions potassium and phosphorus were released instead of taken up at 86 and 189 mg TS-S/L, respectively. This indicates that the sulphide caused a strong inhibition on PAO, where possibly the energy generated (ATP) by the oxidation of PHA was not sufficient thus phosphate was released to provide extra ATP. More severely affected was the growth of PAO as no ammonia consumption (which is frequently associated with growth in EBPR systems) was observed, in neither the direct nor the reversible test.

Long term effects of sulphide on the biological removal of phosphorus: The role of *Thiothrix caldifontis*

During this research, the results gathered up to 20 mg S/L were as expected, as the activities of the anaerobic and aerobic metabolism of *Candidatus* Accumulibacter phosphatis decreased according to an increase in the concentration of sulphide. Above 20 mg S/L, a new organism identified as *Thiothrix caldifontis* proliferated according to an increase in the sulphide concentration. The appearance of *T. caldifontis* was related to the improvement of the biological removal of phosphorus. Nevertheless, the proliferation of this organism caused bulking sludge, which resulted in a reduction of the solids retention time from 20d to 4.6d due to suspended solids in the effluent. Further activity tests and staining technics showed that *T. caldifontis* significantly contributed to the biological removal of phosphorus. Based on Nile blue A, BODYPY, MAR-FISH and mass balance calculations it was possible to conclude that *T. caldifontis* was capable to anaerobically store more than 50 % of the acetate fed. In addition, it was possible to estimate that *T. caldifontis* was capable to at least store 100 mg P/gVSS. Moreover, the sulphur profiles and contrast images suggest that *T. caldifontis* oxidized sulphide into elemental sulphur and stored it as poly-sulphur (Poly-S), to be oxidized into sulphate at a later stage.

Based on the previous results, it is suggested that under anaerobic conditions *T. caldifontis* hydrolyses poly-phosphate to generate energy for the storage of acetate as PHA. In the subsequent aerobic stage, *T. caldifontis* oxidize sulphide into elemental sulphur and store it as Poly-S. The stored Poly-S and PHA are later used to replenish their poly-phosphate storage, to grow and for maintenance purposes. These dual energy pools give them an advantage to grow over other PHA accumulating organisms.

Cooperation between Competibacter sp. and Accumulibacter in denitrification and phosphate removal processes

In this research, it was not possible to observe any significant anoxic phosphorus uptake rate in the enrich culture of PAO I when using nitrate as electron acceptor. On the contrary, the PAO I-GAO culture was capable to anoxically take up phosphorus up to a rate of 8.7 mg PO_4-P/gVSS.h. Moreover, the observed anoxic phosphorus uptake rate when nitrite was used as electron acceptor was not considerably different among the PAO I and PAO I-GAO culture (8.7 ± 0.3 and 9.6 ± 1.8 mgPO_4-P/gVSS.h in the PAO I and PAO I-GAO cultures, respectively). These results suggest that PAO I prefer the use of nitrite to take up phosphorus under anoxic conditions. Thus, it is proposed that GAO might not only compete with PAO for substrate (e.g. acetate) in the anaerobic period, but could also enhanced the anoxic phosphorus removal by reducing nitrate into nitrite.

Absent anoxic activity of PAO I on nitrate under different long-term operational conditions

At any of the operational conditions applied, it was not possible to observe a considerable anoxic phosphorus uptake, where in the best case scenario it reached 13 % of the aerobic phosphorus uptake rate. Moreover, it was possible to observe that under VFA limiting conditions phosphate was anoxically released whereas at higher VFA concentrations a negligible phosphorus uptake was observed. This seems to indicate that the PAO I present in this system could not use nitrate to generate energy and thus PAO need to obtain energy required for maintenance by releasing phosphorus. In line with this observation, the system completely collapsed when the aerobic SRT was reduced below the minimum required for growth. As nitrate was not limiting during anoxic conditions, this indicates that PAO I could not sustain its metabolic conditions using nitrate as electron acceptor. Thus, it seems that PAO I cannot use nitrate to generate energy, whereas is still doubtful the anoxic phosphorus uptake rates observed in past PAO I cultures. Thus, it is proposed that the anoxic phosphorus uptake rate observed in wastewater treatment plants might be due to the respiration of nitrite, which could had been generated by side communities.

7.3. Evaluation and outlook

Effect of electron acceptors in the sulphate reduction process

In this research the aim was to assess if sulphate reducers could proliferate once the sludge is exposed to anoxic and aerobic conditions. Based on these results, it was suggested that a recirculation from the aerobic tank and short anaerobic retention time could hinder the growth of sulphate reducers. However, sulphate reducers should not be seen just as an undesired process as sulphide can be used for metal removal, autotrophic denitrification or even reduction of pathogen (Abdeen et al., 2010; Lewis, 2010; Carmen et al., 2013).

The results obtained in this research could be used to enhance the growth of sulphate reducers in anaerobic tanks. Moreover, it was observed that sulphate reducers capable to use more complex carbon sources were less inhibited by the electron acceptors than sulphate reducers which could only use acetate or propionate. Hence, if the conditions are properly controlled it might be possible to use sulphate reducers to generate acetate and propionate (from the reduction of lactate), which could be used by PAO in order to improve the biological removal of phosphorus.

On the other hand, it has been demonstrated that one of the effects of sulphide in the microbial process it is the deprotonation of dihydrogen sulphide inside the cell membrane (Comeau et al., 1986; Koster et al., 1986). Past research has suggested a similar inhibition pattern by nitrous acid than dihydrogen sulphide on the metabolism of PAOs (Saad et al., 2013). On the other hand, nitrous acid can be used to enhance the hydrolysis of anaerobic digesters, which is normally the limiting process for biogas production (Wang et al., 2013). Thus, as sulphide could be produced in the anaerobic zones of the wastewater treatment plant, it is interesting to asses if it could be used in a similar way that nitrous acid to enhance the hydrolysis in anaerobic digesters or pre-acidification units.

In a different perspective, the sulphate reduction activity could be inhibited in the anaerobic digesters, in this way more carbon could be available for the biogas production. For example in this research it was observed a 0.4h period of no sulphate reduction activity after the biomass was exposed to nitrite. On the contrary, the methane production activity was observed to start immediately after nitrite was not detected in the liquid phase (Borges et al., 2015). Thus, nitrite could be formed in the anoxic tanks of wastewater treatment plants and could be added at different time intervals into the anaerobic digesters. In this way methane production could be favoured over sulphate reduction.

Sulphide effects on the physiology of Candidatus Accumulibacter phosphatis Type I

During this research, it was shown the sulphide strongly inhibited the aerobic metabolism of PAO, even at sulphide concentrations as low as 8 mg H_2S-S/L. Thus, the sulphide potentially formed in the sewage could significantly disturb the biological removal of phosphorus. Although, it was proposed that an extended aerobic phase could alleviate the effect of sulphide this still needs to be confirmed.

Moreover, the effect of sulphide on the aerobic metabolism of PAO could depend on the incubation time. In other words, the time that PAO is exposed to sulphide. Such time could potentially be reduced by the presence of sulphide oxidizing bacteria, which could use the oxygen or nitrate/nitrite present to reduce the sulphide into elemental sulphur. Thus, the ability of PAO to survive to different concentrations of sulphide produced in the sewage could be affected by the presence or absence of bacteria capable to oxidize sulphide.

Another possibility would be to install a sulphide sensor at the start of the plant, the sensor could be connected to a blower which would turn on or off according to the sulphide concentration. Nevertheless, such practice would also reduce the readable biodegradable COD which is necessary for the biological removal of phosphorus and cause oxygen intrusion. Therefore, this hypothesis should be studied in detail.

Finally, it will be interested to design a smart control of iron in order to prevent either high concentrations of sulphide in the influent or high concentration of phosphorus in the effluent. The mathematical expressions presented in this research could be extended to include the iron dosage, and calculate how much iron would need to be dosed in either the influent, effluent or both in order to reach the phosphorus effluent standards.

Long-term effects of sulphide on the biological removal of phosphorus: The role of Thiothrix caldifontis

The potential biological removal of phosphorus by *Thiothrix caldifontis* is the main finding of this research. Nevertheless, the different metabolic pathways are still not completely clear. In order to study this in detail is necessary to enrich a culture of *T. caldifontis*. To do so, it is suggested to operate the bioreactor at an SRT of 4 days, as it was observed that this solids retention time is enough to promote the growth of *Thiothrix*. *Candidatus* Accumulibacter phosphatis could be washed out by alternating carbon and sulphide feeding, as *Ca.* Accumulibacter would be on stress under the presence of sulphide and without PHA is expected that *Ca.* Accumulibacter does not survive. Special attention should be paid to the accumulation of sulphur, as it is hypothesized that *T. caldifontis* would first store elemental sulphur as Poly-S and later use the energy generated by the oxidation of Poly-S into sulphate for the replenishment of Poly-P. In other words, if an excessive amount of Poly-S is stored (due to low SRT or high sulphur load) and this is not oxidized into sulphate, most likely no phosphorus would be stored which potentially could result in the failure of the system.

From a most broad perspective, it has been observed that *T. caldifontis* can reduce nitrate. However, it has not been reported if *T. caldifontis* can store phosphorus under anoxic conditions, which would be interesting to assess. Moreover, in this study it was observed a higher consumption of ammonia most possible due to the mixotrophic growth, so it might be interesting to assess the total nitrogen removal capacity of this organism.

Based on the potential storage polymers of *T. caldifontis* (sulphur and phosphorus) is interesting to design a method to recover these nutrients. For example, due to the bad settleability of the sludge, a membrane bioreactor could be installed after the secondary settler. In this way, the biomass could be separated and the sulphur and phosphorus recovered.

Finally, once the complete metabolism of *T. caldifontis* is understood, would be interesting to design and integrate its metabolic model into the activated sludge model. This model could be used to control the growth not only of *Thiothrix* but could be expanded to include the potential proliferation of filamentous bulking sludge organisms.

Cooperation between Competibacter sp. and Accumulibacter in denitrification and phosphate removal processes

In this study, it seems clear that PAO I could not generate energy from the respiration of nitrate as efficient as with oxygen or even nitrite. For this reason, it might be more attractive to integrate an anoxic phosphorus removal based on nitrite reduction than nitrate. This might be attractive to treat effluents high in ammonia (e.g. industrial, anaerobic digesters) as the oxygen demand can be reduced considerable by the partial oxidation of ammonia into nitrite. Moreover, the possible inhibition on PAOs due to the formation of free nitrous acid could be reduced by coupling the PAOs with other microbes such as Anammox. Thus, it might be important to study the interaction and possibility of cooperation between PAO and Anammox in the treatment of high ammonia and low COD effluents.

Absent anoxic activity of PAO I on nitrate under different long-term operational conditions

In the past, it was suggested that a PAO capable to use nitrate to generate energy to take up phosphorus existed. It was believed that PAO I was such PAO. Nevertheless, these studies showed that PAO I is not capable to survive solely under anoxic conditions, which suggest that PAO I cannot generate enough (or none) energy from the respiration of nitrate to sustain its metabolic activities. Nevertheless, in this research we operate the reactor during long-term (months) under anaerobic-anoxic-oxic conditions, thus the oxic conditions could limit the growth of a true DPAO. Thus, it might be interesting to try to enrich a PAO under anaerobic-anoxic conditions. In order to prevent the proliferation of GAO or other PHA accumulating organism, the reactor could be fed continuously with acetate and nitrate (Kuba et al., 1993; Tu et al., 2013),operate at a high pH (Filipe et al., 2001) and low temperature (Lopez-Vazquez et al., 2008).

Once such PAO is identified, it might be useful to determinate its kinetics, half saturation constant, gas production and growth so it can be added to a model. The model then can be used to enhance the anoxic phosphorus removal and reduction of carbon demand during denitrification of wastewater treatment plant.

8

References

A

Abdeen, S., Di, W., Hui, L., Chen, G.-H., van Loosdrecht, M.C.M., 2010. Fecal coliform removal in a sulfate reduction, autotrophic denitrification and nitrification integrated (SANI) process for saline sewage treatment. Water science and technology : a journal of the International Association on Water Pollution Research 62 (11), 2564–70.

Ahn, J., Daidou, T., Tsuneda, S., Hirata, A., 2001a. Selection and dominance mechanisms of denitrifying phosphate-accumulating organisms in biological phosphate removal process. Biotechnology letters 2005–2008.

Ahn, J., Daidou, T., Tsuneda, S., Hirata, A., 2002. Characterization of denitrifying phosphate-accumulating organisms cultivated under different electron acceptor conditions using polymerase chain reaction-denaturing gradient gel electrophoresis assay. Water research 36 (2), 403–12.

Ahn, J., Daidou, T., Tsuneda, S., Hirata, a, 2001b. Metabolic behavior of denitrifying phosphate-accumulating organisms under nitrate and nitrite electron acceptor conditions. Journal of bioscience and bioengineering 92 (5), 442–6.

Albertsen, M., Karst, S.M., Ziegler, A.S., Kirkegaard, R.H., Nielsen, P.H., 2015. Back to basics - The influence of DNA extraction and primer choice on phylogenetic analysis of activated sludge communities. PLoS ONE 10 (7), 1–15.

Amman, R.I., 1995. In situ identification of micro-organisms by whole cell hybridization with rRNA-targeted nucleic acid probes. In: Ackkermans, A., van Elsas, J., de Bruijn, F. (Eds.), Molecular Microbial Ecology Manual. Klower Academy Publications, Dordrecht, Holland.

APHA, AWWA, WEF, 2005. Standard Methods for the Examination of Water and Wastewater, 22th ed. American Water Works Assn.

Baetens, D., Weemaes, M., Hosten, L., de Vos, P., Vanrolleghem, P., 2001. Enhanced Biological Phosphorus Removal: Competition and symbiosis between SRBs and PAOs on lactate/acetate feed (1987), 1994–1997.

Barat, R., Montoya, T., Seco, a, Ferrer, J., 2005. The role of potassium, magnesium and calcium in the Enhanced Biological Phosphorus Removal treatment plants. Environmental technology 26 (9), 983–992.

Barnard, J.L., 1975. Biological nutrient removal without the addition of chemicals. Water Research 9 (5–6), 485–490.

Barton, L.L., Hamilton, W.A., 2007. Sulphate-Reducing Bacteria: Environmental and Engineered Systems. Cambridge.

Bassin, J.P., Pronk, M., Muyzer, G., Kleerebezem, R., Dezotti, M., van Loosdrecht, M.C.M., 2011. Effect of elevated salt concentrations on the aerobic granular sludge process: linking microbial activity with microbial community structure. Applied and environmental microbiology 77 (22), 7942–53.

Baumgartner, L.K., Reid, R.P., Dupraz, C., Decho, a. W., Buckley, D.H., Spear, J.R., Przekop, K.M., Visscher, P.T., 2006. Sulfate reducing bacteria in microbial mats: Changing paradigms, new discoveries. Sedimentary Geology 185 (3–4), 131–145.

Bedzyk, L., Wang, T., Ye, R.W., 1999. The periplasmic nitrate reductase in Pseudomonas sp. strain G-179 catalyzes the first step of denitrification. Journal of Bacteriology 181 (9), 2802–2806.

Bejarano Ortiz, D.I., Thalasso, F., Cuervo López, F.D.M., Texier, A.C., 2013. Inhibitory effect of sulfide on the nitrifying respiratory process. Journal of Chemical Technology and Biotechnology 88 (October 2012), 1344–1349.

Bell, L.C., Page, M.D., Berks, B.C., Richardson, D.J., Ferguson, S.J., 1993. Insertion of transposon Tn5 into a structural gene of the membrane-bound nitrate reductase of Thiosphaera pantotropha results in anaerobic overexpression of periplasmic nitrate reductase activity. Journal of general microbiology 139 (12), 3205–14.

Bentzen, G., Smith, T.A., Bennett, D., Webster, N.J., Reinholt, F., Sletholt, E., Hobson, J., 1995. Controlled dosing of nitrate for prevention of H2S in a sewer network and the effects on the subsequent treatment processes. Water Science & Technology 31 (7), 293–302.

Berg, J.S., Schwedt, A., Kreutzmann, A.C., Kuypers, M.M.M., Milucka, J., 2014. Polysulfides as intermediates in the oxidation of sulfide to sulfate by Beggiatoa spp. Applied and Environmental Microbiology 80 (2), 629–636.

Bond, P.L., Hugenholtz, P., Keller, J., Blackall, L.L., 1995. Bacterial community structures of phosphate-removing and non-phosphate-removing activated sludges from sequencing batch reactors. Applied and environmental microbiology 61 (5), 1910–1916.

Borges, L.I., López-Vazquez, C.M., García, H., Van Lier, J.B., 2015. Nitrite reduction and methanogenesis in a single-stage UASB reactor. Water Science and Technology 72 (12), 2236–2242.

van den Brand, T.P.H., Roest, K., Brdjanovic, D., Chen, G.H., van Loosdrecht, M.C.M., 2014a. Temperature effect on acetate and propionate consumption by sulfate-reducing bacteria in saline wastewater. Applied microbiology and biotechnology 98 (9), 4245–55.

van den Brand, T.P.H., Roest, K., Brdjanovic, D., Chen, G.H., van Loosdrecht, M.C.M., 2014b. Influence of acetate and propionate on sulphate reducing bacteria activity. Journal of applied microbiology 31 (0).

van den Brand, T.P.H., Roest, K., Brdjanovic, D., Chen, G.H., van Loosdrecht, M.C.M., 2014c. Influence of acetate and propionate on sulphate-reducing bacteria activity. Journal of Applied Microbiology 117, 1839–1847.

van den Brand, T.P.H., Roest, K., Chen, G.H., Brdjanovic, D., van Loosdrecht, M.C.M.,

2015. Occurrence and activity of sulphate reducing bacteria in aerobic activated sludge systems. World Journal of Microbiology and Biotechnology 31 (3), 507–516.

Bratby, J., 2016. Coagulation and Flocculation in Water and Wastewater Treatment, Third edit. ed. IWA publishing.

Brdjanovic, D., Hooijmans, C.M., van Loosdrecht, M.C.M., Alaerts, G.J., Heijnen, J.J., 1996. The dynamic effects of potassium limitation on biological phosphorus removal. Water Research 30 (10), 2323–2328.

Brdjanovic, D., Van Loosdrecht, M.C.M., Hooijmans, C.M., Alaerts, G.J., Heijnen, J.J., 1998. Minimal aerobic sludge retention time in biological phosphorus removal systems. Biotechnology and Bioengineering 60 (3), 326–332.

Brock, J., Rhiel, E., Beutler, M., Salman, V., Schulz-Vogt, H.N., 2012. Unusual polyphosphate inclusions observed in a marine Beggiatoa strain. Antonie van Leeuwenhoek, International Journal of General and Molecular Microbiology 101 (2), 347–357.

Brock, J., Schulz-Vogt, H.N., 2011. Sulfide induces phosphate release from polyphosphate in cultures of a marine Beggiatoa strain. The ISME journal 5, 497–506.

Brown, A.D., 1990. Microbial water stress physiology; Principles and perspectives. WILEY, John Wiley & Sons, Inc.

Camejo, P.Y., Owen, B.R., Martirano, J., Ma, J., Kapoor, V., Santodomingo, J., McMahon, K.D., Noguera, D.R., 2016. Candidatus Accumulibacter phosphatis clades enriched under cyclic anaerobic and microaerobic conditions simultaneously use different electron acceptors. Water Research 102, 125–137.

Cao, J., Zhang, G., Mao, Z.S., Li, Y., Fang, Z., Yang, C., 2012. Influence of electron donors on the growth and activity of sulfate-reducing bacteria. International Journal of Mineral Processing 106–109, 58–64.

Cardenas, J., Kerber, N.L., 1982. Nitrate reductase from Rhodopseudomonas sphaeroides. Journal of Bacteriology 150 (3), 1091–1097.

Carmen, F., Anuska, M.-C., Luis, C.J., Ramón, M., 2013. Post-treatment of fish canning effluents by sequential nitrification and autotrophic denitrification processes. Process Biochemistry 48 (9), 1368–1374.

Carvalho, G., Lemos, P.C., Oehmen, A., Reis, M.A.., 2007. Denitrifying phosphorus removal: linking the process performance with the microbial community structure. Water Research 41 (19), 4383–4396.

Cech, J.S., Hartman, P., Wanner, J., 1993. Competition between PolyP and Non-PolyP Bacteria in an Enhanced Phosphate Removal System. Water Environment Research 65 (5), 690–692.

Chaime, J., 2004. World Population to 2300. United Nations Department of Economic and

Social Affairs/Population Division.

Chen, Y., Chen, H., Zheng, X., Mu, H., 2012. The impacts of silver nanoparticles and silver ions on wastewater biological phosphorous removal and the mechanisms. Journal of Hazardous Materials 239–240, 88–94.

Chen, Y., Cheng, J.J., Creamer, K.S., 2008. Inhibition of anaerobic digestion process: a review. Bioresource technology 99 (10), 4044–64.

Chernousova, E., Gridneva, E., Grabovich, M., Dubinina, G., Akimov, V., Rossetti, S., Kuever, J., 2009. Thiothrix caldifontis sp. nov. and Thiothrix lacustris sp. nov., gammaproteobacteria isolated from sulfide springs. International Journal of Systematic and Evolutionary Microbiology 59 (12), 3128–3135.

Choi, E., Rim, J.M., 1991. Competition and inhibition of sulfate reducers and methane producers in anaerobic treatment. Water Science and Technology 23 (7–9), 1259–1264.

Comeau, Y., Hall, K., Hancock, R., Oldham, W., 1986. Biochemical model for enhanced biological phosphorus removal. Water Research 20 (12), 1511–1521.

Cooper, M.S., Hardin, W.R., Petersen, T.W., Cattolico, R.A., 2010. Visualizing "green oil" in live algal cells. Journal of Bioscience and Bioengineering 109 (2), 198–201.

Crocetti, G.R., Hugenholtz, P., Bond, P.L., Schuler, a, Keller, J., Jenkins, D., Blackall, L.L., 2000. Identification of polyphosphate-accumulating organisms and design of 16S rRNA-directed probes for their detection and quantitation. Applied and environmental microbiology 66 (3), 1175–82.

Cypionka, H., 1994. Novel metabolic capacities of sulfate-reducing bacteria, and their activities in microbial mats. In: Microbial Mats. Springer Berlin Heidelberg, Berlin, Heidelberg.

Cypionka, H., Widdel, F., Pfennig, N., 1985. Survival of sulfate-reducing bacteria after oxygen stress, and growth in sulfate-free oxygen-sulfide gradients. FEMS Microbiology Letters 31 (1), 39–45.

Daigger, G.T., Hodgkinson, A., Aquilina, S., Fries, M.K., 2015. Development and Implementation of a Novel Sulfur Removal Process from H2S Containing Wastewaters. Water Environment Research 87 (7), 618–625.

Dale, B.E., White, D.H., 1983. Ionic strength: A neglected variable in enzyme technology. Enzyme and Microbial Technology 5 (3), 227–229.

Dar, S. a., Kleerebezem, R., Stams, a. J.M., Kuenen, J.G., Muyzer, G., 2008. Competition and coexistence of sulfate-reducing bacteria, acetogens and methanogens in a lab-scale anaerobic bioreactor as affected by changing substrate to sulfate ratio. Applied Microbiology and Biotechnology 78, 1045–1055.

Dilling, Waltraud, Cypionka, H., 1990. Aerobic respiration in sulfate-reducing bacteria. FEMS Microbiology Letters 71, 123–128.

Dolla, A., Fournier, M., Dermoun, Z., 2006. Oxygen defense in sulfate-reducing bacteria. Journal of biotechnology 126 (1), 87–100.

EEA, 2005. Source apportionmetn of nitrogen and phosphorous inputs into the aquatic enviroment. Copenhagen, Denmark.

Ekama, G.A., Wentzel, M.C., 1999. Difficulties and developments in biological nutrient removal technology and modelling. Water Science and Technology 39 (6), 1–11.

Ekama, G. a., Wilsenach, J. a., Chen, G.H., 2011. Saline sewage treatment and source separation of urine for more sustainable urban water management. Water Science & Technology 64 (6), 1307.

van de Ende, F.P., Meier, J., van Gemerden, H., 1997. Syntrophic growth of sulfate-reducing bacteria and colorless sulfur bacteria during oxygen limitation. Microbial ecology 23, 65–80.

Fang, J., Sun, P. De, Xu, S.J., Luo, T., Lou, J.Q., Han, J.Y., Song, Y.Q., 2012. Impact of Cr(VI) on P removal performance in enhanced biological phosphorus removal (EBPR) system based on the anaerobic and aerobic metabolism. Bioresource Technology 121, 379–385.

Filipe, C.D., Daigger, G.T., Grady, C.P., 2001. pH as a key factor in the competition between glycogen-accumulating organisms and phosphorus-accumulating organisms. Water environment research : a research publication of the Water Environment Federation 73 (2), 223–32.

Flowers, J.J., He, S., Carvalho, G., Peterson, S.B., Lopez, C., Yilmaz, S., Zilles, J.L., Morgenroth, E., Lemos, P.C., M, M. a, Crespo, M.T.B., Noguera, D.R., Mcmahon, K.D., 2008. Ecological Differentiation of Accumulibacter in EBPR Reactors. Environment 2008 (17), 31–42.

Flowers, J.J., He, S., Malfatti, S., del Rio, T.G., Tringe, S.G., Hugenholtz, P., McMahon, K.D., 2013. Comparative genomics of two "Candidatus Accumulibacter" clades performing biological phosphorus removal. The ISME journal 7 (12), 2301–14.

Flowers, J.J., He, S., Yilmaz, S., Noguera, D.R., McMahon, K.D., 2009. Denitrification capabilities of two biological phosphorus removal sludges dominated by different "Candidatus Accumulibacter" clades. Environmental microbiology reports 1 (6), 583–588.

Fuhs, G.W., Chen, M., 1975. Microbiological basis of phosphate removal in the activated sludge process for the treatment of wastewater. Microbial Ecology 2 (2), 119–138.

Galinski, E., Trüper, H., Reviews, F.M., Societies, E.M., Friedrich-wilhehns-, R., Auee, M., 1994. Microbial behaviour in salt-stressed ecosystems. FEMS Microbiology Reviews 15, 95–108.

García De Lomas, J., Corzo, A., Gonzalez, J.M., Andrades, J. a., Iglesias, E., Montero, M.J.,

2006. Nitrate promotes biological oxidation of sulfide in wastewaters: Experiment at plant-scale. Biotechnology and Bioengineering 93 (4), 801–811.

García Martín, H., Ivanova, N., Kunin, V., Warnecke, F., Barry, K.W., McHardy, A.C., Yeates, C., He, S., Salamov, A. a, Szeto, E., Dalin, E., Putnam, N.H., Shapiro, H.J., Pangilinan, J.L., Rigoutsos, I., Kyrpides, N.C., Blackall, L.L., McMahon, K.D., Hugenholtz, P., 2006. Metagenomic analysis of two enhanced biological phosphorus removal (EBPR) sludge communities. Nature biotechnology 24 (10), 1263–9.

Ginestet, P., Nicol, R., Holst, T., Lebossé, X., 2015. Evidence for sulfide associated autotrophic biological phosphorus removal in a full scale wastewater treatment plant. WA Nutrient Removal and Recovery 2015: moving innovation into practice.

Gonzalez-Gil, G., Holliger, C., 2011. Dynamics of microbial community structure of and enhanced biological phosphorus removal by aerobic granules cultivated on propionate or acetate. Applied and Environmental Microbiology 77 (22), 8041–8051.

Greene, E. a, Hubert, C., Nemati, M., Jenneman, G.E., Voordouw, G., 2003. Nitrite reductase activity of sulphate-reducing bacteria prevents their inhibition by nitrate-reducing, sulphide-oxidizing bacteria. Environmental microbiology 5 (7), 607–17.

Guerrero, J., Tayà, C., Guisasola, A., Baeza, J. a., 2012. Understanding the detrimental effect of nitrate presence on EBPR systems: Effect of the plant configuration. Journal of Chemical Technology and Biotechnology 87 (10), 1508–1511.

Guisasola, A., Qurie, M., Vargas, M. del M., Casas, C., Baeza, J.A., 2009. Failure of an enriched nitrite-DPAO population to use nitrate as an electron acceptor. Process Biochemistry 44, 689–695.

Guo, G., Wu, D., Hao, T., Mackey, H.R., Wei, L., Wang, H., Chen, G., 2016a. Functional bacteria and process metabolism of the Denitrifying Sulfur conversion-associated Enhanced Biological Phosphorus Removal (DS-EBPR) system: An investigation by operating the system from deterioration to restoration. Water Research 95, 289–299.

Guo, G., Wu, D., Hao, T., Mackey, H.R., Wei, L., Wang, H., Chen, G., 2016b. Functional bacteria and process metabolism of the Denitrifying Sulfur conversion-associated Enhanced Biological Phosphorus Removal (DS-EBPR) system: An investigation by operating the system from deterioration to restoration. Water Research.

Halkjær, N., Daims, H., Lemmer, H., 2009. FISH Handbook for Biological Wastewater Treatment. IWA publishing.

Hao, T., Xiang, P., Mackey, H.R., Chi, K., Lu, H., Chui, H., van Loosdrecht, M.C.M., Chen, G.-H., 2014. A Review of Biological Sulfate Conversions in Wastewater Treatment. Water Research 65, 1–21.

He, S., Gall, D.L., McMahon, K.D., 2007. "Candidatus Accumulibacter" population structure in enhanced biological phosphorus removal sludges as revealed by polyphosphate kinase genes. Applied and environmental microbiology 73 (18), 5865–74.

He, S., Gu, A.Z., McMahon, K.D., 2005. The role of Rhodocyclus-like organisms in biological phosphorus removal: factors influencing population structure and activity. Proceedings of the Water Environment Federation 2005 (14), 1999–2011.

Henze, M., van Loosdrecht, M.C.M., Ekama, G.A., Brdjanovic, D., 2008. Biological Wastewater Treatment-Principles, Modelling and Design, 1st ed. IWA publishing.

Hesselmann, R.P., Werlen, C., Hahn, D., van der Meer, J.R., Zehnder, a J., 1999. Enrichment, phylogenetic analysis and detection of a bacterium that performs enhanced biological phosphate removal in activated sludge. Systematic and applied microbiology 22 (3), 454–65.

Hubert, C., Nemati, M., Jenneman, G., Voordouw, G., 2005. Corrosion risk associated with microbial souring control using nitrate or nitrite. Applied Microbiology and Biotechnology 68 (2), 272–282.

Ichihashi, O., Satoh, H., Mino, T., 2006a. Effect of soluble microbial products on microbial metabolisms related to nutrient removal. Water Research 40 (8), 1627–1633.

Ichihashi, O., Satoh, H., Mino, T., 2006b. Effect of soluble microbial products on microbial metabolisms related to nutrient removal. Water research 40 (8), 1627–33.

Isa, Z., Grusenmeyer, S., Verstraete, W., 1986. Sulfate reduction relative to methane production in high-rate anaerobic digestion: microbiological aspects. Applied and environmental microbiology 51 (3), 580–7.

Jeong Myeong, K., Hyo Jung, L., Dae Sung, L., Che Ok, J., 2013. Characterization of the denitrification-associated phosphorus uptake properties of "Candidatus Accumulibacter phosphatis" clades in sludge subjected to enhanced. Applied and environmental microbiology 79 (6), 1969–79.

Jianlong, W., Ning, Y., 2004. Partial nitrification under limited dissolved oxygen conditions. Process Biochemistry 39 (10), 1223–1229.

Jin, R.C., Yang, G.F., Zhang, Q.Q., Ma, C., Yu, J.J., Xing, B.S., 2013. The effect of sulfide inhibition on the ANAMMOX process. Water Research 47 (3), 1459–1469.

Kanagawa, T., Kamagata, Y., Aruga, S., Kohno, T., Horn, M., Wagner, M., 2000. Phylogenetic Analysis of and Oligonucleotide Probe Development for Eikelboom Type 021N Filamentous Bacteria Isolated from Bulking Activated Sludge Phylogenetic Analysis of and Oligonucleotide Probe Development for Eikelboom Type 021N Filamentous Bacteria. Applied and Environmental Microbiology 66 (11), 5043–5052.

Karagiannis, I.C., Soldatos, P.G., 2008. Water desalination cost literature: review and assessment. Desalination 223 (1–3), 448–456.

Kerrn-Jespersen, J.P., Henze, M., 1993. Biological phosphorus uptake under anoxic and aerobic conditions. Water Research 27 (4), 617–624.

Kjeldsen, K., Joulian, C., Ingvorsen, K., 2004. Oxygen tolerance of sulfate-reducing bacteria in activated sludge. Environmental science & Technology 38, 2038–2043.

Kjeldsen, K.U., Joulian, C., Ingvorsen, K., 2005. Effects of oxygen exposure on respiratory activities of Desulfovibrio desulfuricans strain DvO1 isolated from activated sludge. FEMS microbiology ecology 53 (2), 275–84.

Kleerebezem, R., Mendezà, R., 2002. Autotrophic denitrification for combined hydrogen sulfide removal from biogas and post-denitrification. Water Science and Technology 45 (10), 349 LP-356.

Kong, Y., Nielsen, J.L., Nielsen, P.H., 2005. Identity and ecophysiology of uncultured actinobacterial polyphosphate-accumulating organisms in full-scale enhanced biological phosphorus removal plants. Applied and Environmental Microbiology 71 (7), 4076–4085.

Kong, Y., Ong, S., Ng, W., Liu, W., 2002. Diversity and distribution of a deeply branched novel proteobacterial group found in anaerobic–aerobic activated sludge processes. Environmental Microbiology 4 (11), 753–757.

Kong, Y., Xia, Y., Nielsen, J.L., Nielsen, P.H., 2006. Ecophysiology of a group of uncultured Gammaproteobacterial glycogen-accumulating organisms in full-scale enhanced biological phosphorus removal wastewater treatment plants. Environmental microbiology 8 (3), 479–89.

Koster, I., Rinzema, A., Devegt, A., Lettinga, G., 1986. Sulfide inhibition of the methanogenic activity of granular sludge at various pH-levels. Water Research.

Kristiansen, R., Nguyen, H.T.T., Saunders, A.M., Nielsen, J.L., Wimmer, R., Le, V.Q., McIlroy, S.J., Petrovski, S., Seviour, R.J., Calteau, A., Nielsen, K.L., Nielsen, P.H., 2013. A metabolic model for members of the genus Tetrasphaera involved in enhanced biological phosphorus removal. The ISME journal 7 (3), 543–54.

Kuba, T., Loosdrecht, M. Van, Heijnen, J., 1996a. Phosphorus and nitrogen removal with minimal COD requirement by integration of denitrifying dephosphatation and nitrification in a two-sludge system. Water research 1354 (96), 1702–1710.

Kuba, T., Van Loosdrecht, M.C.M., Brandse, F. a., Heijnen, J.J., 1997a. Occurrence of denitrifying phosphorus removing bacteria in modified UCT-type wastewater treatment plants. Water Research 31 (4), 777–786.

Kuba, T., van Loosdrecht, M.C.M., Heijnen, J.J., 1997b. Biological dephosphatation by activated sludge under denitrifying conditions: pH influence and occurrence of denitrifying dephosphatation in a full-scale waste water treatment plant. Water Science & Technology 36 (12), 75–82.

Kuba, T., Murnleitner, E., van Loosdrecht, M.C., Heijnen, J.J., 1996b. A metabolic model for biological phosphorus removal by denitrifying organisms. Biotechnology and bioengineering 52 (6), 685–95.

Kuba, T., Smolders, G., van Loosdrecht, M.C.M., Heijnen, J.J., 1993. Biological phosphorus removal from wastewater by anaerobic-anoxic sequencing batch reactor. Water Science & Technology 27 (5), 241–252.

Kuenen, J.G., Beudeker, R.F., 1982. Microbiology of thiobacilli and other sulphur-oxidizing autotrophs, mixotrophs and heterotrophs. Philosophical transactions of the Royal Society of London. Series B, Biological sciences 298, 473–497.

Kulakovskaya, T. V, Lichko, L.P., Ryazanova, L.P., 2014. Diversity of Phosphorus Reserves in Microorganisms. Biochemestry 79 (13), 1602–1614.

Kundu, P., Pramanik, A., Dasgupta, A., Mukherjee, S., Mukherjee, J., 2014. Simultaneous heterotrophic nitrification and aerobic denitrification by Chryseobacterium sp. R31 isolated from abattoir wastewater. BioMed Research International 2014.

Lanham, A., Moita, R., Lemos, P., Reis, M., 2011. Long-term operation of a reactor enriched in Accumulibacter clade I DPAOs: performance with nitrate, nitrite and oxygen. Water Science & … 352–359.

Lau, G.N., Sharma, K.R., Chen, G.H., Loosdrecht, M.C.M. va., 2006. Integration of sulphate reduction, autotrophic denitrification and nitrification to achieve low-cost excess sludge minimisation for Hong Kong sewage. Water Science & Technology 53 (3), 227.

Lee, C., Yu, C., 1997. Conservation of water resources- use of sea water for flushing in Hong Kong. Aqua- Journal of Water Supply 46, 202–209.

Lee, H., Yun, Z., 2014. Comparison of biochemical characteristics between PAO and DPAO sludges. Journal of environmental sciences (China) 26 (6), 1340–7.

Lemaire, R., Meyer, R., Taske, A., Crocetti, G.R., Keller, J., Yuan, Z., 2006. Identifying causes for N2O accumulation in a lab-scale sequencing batch reactor performing simultaneous nitrification, denitrification and phosphorus removal. Journal of biotechnology 122 (1), 62–72.

Lens, P., Poorter, M. De, 1995. Sulfate reducing and methane producing bacteria in aerobic wastewater treatment systems. Water Research 29 (3), 871–880.

Lens, P., Vallero, M., Esposito, G., Zandvoort, M., 2002. Perspectives of sulfate reducing bioreactors in environmental biotechnology. Reviews in Environmental Science and Biotechnology 1, 311–325.

Lens, P.N., Kuenen, J.G., 2001. The biological sulfur cycle: novel opportunities for environmental biotechnology. Water science and technology : a journal of the International Association on Water Pollution Research 44, 57–66.

Lewis, A.E., 2010. Review of metal sulphide precipitation. Hydrometallurgy 104 (2), 222–234.

Liamleam, W., Annachhatre, A.P., 2007. Electron donors for biological sulfate reduction. Biotechnology advances 25 (5), 452–63.

Londry, K., Suflita, J., 1999. Use of nitrate to control sulfide generation by sulfate-reducing bacteria associated with oily waste. Journal of industrial microbiology & biotechnology 22 (6), 582–589.

Lopez-Vazquez, C.M., Hooijmans, C.M., Chen, G., Loosdrecht, M.C.M. Van, Brdjanovic, D., 2009a. Use of saline water in sanitation : change of paradigm in water resources management in urban environments 31, 1–9.

Lopez-Vazquez, C.M., Oehmen, A., Hooijmans, C.M., Brdjanovic, D., Gijzen, H.J., Yuan, Z., van Loosdrecht, M.C.M., 2009b. Modeling the PAO-GAO competition: effects of carbon source, pH and temperature. Water research 43 (2), 450–62.

Lopez-Vazquez, C.M., Song, Y.-I., Hooijmans, C.M., Brdjanovic, D., Moussa, M.S., Gijzen, H.J., van Loosdrecht, M.C.M., 2008. Temperature effects on the aerobic metabolism of glycogen-accumulating organisms. Biotechnology and bioengineering 101 (2), 295–306.

Lu, H., Wu, D., Tang, D.T.W., Chen, G.H., van Loosdrecht, M.C.M., Ekama, G., 2011. Pilot scale evaluation of SANI ® process for sludge minimization and greenhouse gas reduction in saline sewage treatment. Water Science & Technology 63 (10), 2149.

Macalady, J.L., Lyon, E.H., Koffman, B., Albertson, L.K., Meyer, K., Galdenzi, S., Mariani, S., 2006. Dominant microbial populations in limestone-corroding stream biofilms, Frasassi cave system, Italy. Applied and Environmental Microbiology 72 (8), 5596–5609.

Maillacheruvu, K., Parkin, G., 1996. Kinetics of growth, substrate utilization and sulfide toxicity for propionate, acetate, and hydrogen utilizers in anaerobic systems. Water environment research 68 (7), 1099–1106.

Maillacheruvu, K., Parkin, G., Peng, C., 1993. Sulfide toxicity in anaerobic systems fed sulfate and various organics. Water Environment Federation 65 (2), 100–109.

Marzluf, G.A., Reddy, C.A., Beveridge, T.J., Schmidt, T.M., Snyder, L.R., Breznak, J.A. (Eds.), 2007. Methods for General and Molecular Microbiology, Third Edition. American Society of Microbiology.

McIlroy, S.J., Albertsen, M., Andresen, E.K., Saunders, A.M., Kristiansen, R., Stokholm-Bjerregaard, M., Nielsen, K.L., Nielsen, P.H., 2014. "Candidatus Competibacter"-lineage genomes retrieved from metagenomes reveal functional metabolic diversity. The ISME journal 8 (3), 613–24.

McIlroy, S.J., Saunders, A.M., Albertsen, M., Nierychlo, M., McIlroy, B., Hansen, A.A., Karst, S.M., Nielsen, J.L., Nielsen, P.H., 2015. MiDAS: The field guide to the microbes of activated sludge. Database 2015 (2), 1–8.

McMahon, K.D., Dojka, M.A., Pace, N.R., Jenkins, D., Keasling, J.D., 2002. Polyphosphate Kinase from Activated Sludge Performing Enhanced Biological Phosphorus Removal Polyphosphate Kinase from Activated Sludge Performing Enhanced Biological Phosphorus Removal †. Applied and Environmental Microbiology 68 (10), 4971–4978.

McMahon, K.D., Yilmaz, S., He, S., Gall, D.L., Jenkins, D., Keasling, J.D., 2007. Polyphosphate kinase genes from full-scale activated sludge plants. Applied microbiology and biotechnology 77 (1), 167–73.

Mergaert, J., Cnockaert, M.C., Swings, J., 2003. Thermomonas fusca sp. nov. and Thermomonas brevis sp. nov., two mesophilic species isolated from a denitrification reactor with poly(E-caprolactone) plastic granules as fixed bed, and emended description of the genus Thermomonas. International Journal of Systematic and Evolutionary Microbiology 53 (6), 1961–1966.

Metcalf & Eddy, Tchobanoglous, G., Burton, F.L., Stensel, H.D., 2003. Wastewater Engineering: Treatment and Reuse, 4th ed. McGraw-Hill Professional.

Meyer, R.L., Saunders, A.M., Blackall, L.L., 2006. Putative glycogen-accumulating organisms belonging to the Alphaproteobacteria identified through rRNA-based stable isotope probing. Microbiology 152 (2), 419–429.

Mino, T., Loosdrecht, M. Van, Heijnen, J., 1998. Microbiology and biochemistry of the enhanced biological phosphate removal process. Water research 32 (11).

Mino, T., Tsuzuki, Y., Matsuo, T., 1987. Effect of phosphorus accumulation on acetate metabolism in the biological phosphorus removal process. In: Biological Phosphate Removal from Wastewaters. Pergamon Press, Oxford.

Mohanakrishnan, J., Gutierrez, O., Meyer, R.L., Yuan, Z., 2008. Nitrite effectively inhibits sulfide and methane production in a laboratory scale sewer reactor. Water research 42 (14), 3961–71.

Mohanakrishnan, J., Gutierrez, O., Sharma, K.R., Guisasola, a, Werner, U., Meyer, R.L., Keller, J., Yuan, Z., 2009. Impact of nitrate addition on biofilm properties and activities in rising main sewers. Water research 43 (17), 4225–37.

Moreno-Vivián, C., Cabello, P., Blasco, R., Castillo, F., Cabello, N., Marti, M., Moreno-vivia, C., 1999. Prokaryotic Nitrate Reduction : Molecular Properties and Functional Distinction among Bacterial Nitrate Reductases. Journal of Bacteriology 181 (21), 6573–6584.

Mulder, A., van de Graaf, A., Robertson, L.A., Kuenen, J.G., 1995. Anaerobic ammonium oxidation discovered in a denitrifying fluidized bed reactor. FEMS Microbiology Ecology 16 (3), 177–184.

Muyzer, G., Stams, A.J.M., 2008. The ecology and biotechnology of sulphate-reducing bacteria. Nature reviews. Microbiology 6 (6), 441–54.

Nemati, M., Mazutinec, T.J., Jenneman, G.E., Voordouw, G., 2001. Control of biogenic H(2)S production with nitrite and molybdate. Journal of industrial microbiology & biotechnology 26 (6), 350–5.

NEN 6472, 1983. Water Forometrische bepaling van het gehalte aan ammonium.

Nguyen, H.T.T., Le, V.Q., Hansen, A.A., Nielsen, J.L., Nielsen, P.H., 2011. High diversity and abundance of putative polyphosphate-accumulating Tetrasphaera-related bacteria in activated sludge systems. FEMS microbiology ecology 76 (2), 256–67.

Nielsen, A.H., Vollertsen, J., Jensen, H.S., Madsen, H.I., Hvitved-Jacobsen, T., 2008. Aerobic and anaerobic transformations of sulfide in a sewer system--field study and model simulations. Water environment research : a research publication of the Water Environment Federation 80 (1), 16–25.

Nielsen, J.L., Nielsen, P.H., 2005. Advances in microscopy: Microautoradiography of single cells. Methods in Enzymology 397 (2004), 237–256.

Nielsen, P.H., Daims, H., Lemmer, H., Arslan-Alaton, I., Olmez-Hanci, T., 2009. FISH Handbook for Biological Wastewater Treatment. IWA Publishing.

Nielsen, P.H., Muro, M.A. de, Nielsen, J.L., 2000. Studies on the *in situ* physiology of *Thiothrix* spp. present in activated sludge. Environmental Microbiology 2 (4), 389–398.

Oehmen, A., Carvalho, G., Freitas, F., Reis, M.A., 2010a. Assessing the abundance and activity of denitrifying polyphosphate accumulating organisms through molecular and chemical techniques. Water Science & Technology 61 (8), 2061.

Oehmen, A., Carvalho, G., Lopez-Vazquez, C.., van Loosdrecht, M.C.., Reis, M.A.M., 2010b. Incorporating microbial ecology into the metabolic modelling of polyphosphate accumulating organisms and glycogen accumulating organisms. Water research 44 (17), 4992–5004.

Oehmen, A., Lemos, P.C., Carvalho, G., Yuan, Z., Keller, J., Blackall, L.L., Reis, M. a M., 2007. Advances in enhanced biological phosphorus removal: from micro to macro scale. Water research 41 (11), 2271–300.

Oehmen, A., Lopez-Vazquez, C.M., Carvalho, G., Reis, M.A.M., van Loosdrecht, M.C.M., 2010c. Modelling the population dynamics and metabolic diversity of organisms relevant in anaerobic/anoxic/aerobic enhanced biological phosphorus removal processes. Water research 44 (15), 4473–86.

Oehmen, A., Yuan, Z., Blackall, L.L., Keller, J., 2004. Short-term effects of carbon source on the competition of polyphosphate accumulating organisms and glycogen accumulating organisms. Water science and technology : a journal of the International Association on Water Pollution Research 50 (10), 139–44.

Oehmen, A., Yuan, Z., Blackall, L.L., Keller, J., 2005. Comparison of acetate and propionate uptake by polyphosphate accumulating organisms and glycogen accumulating organisms. Biotechnology and bioengineering 91 (2), 162–8.

Okabe, S., Ito, T., Sugita, K., Satoh, H., 2005. Succession of internal sulfur cycles and sulfur-oxidizing bacterial communities in microaerophilic wastewater biofilms. Applied and Enviromental Microbiology 71 (5), 2520–2529.

Oyekola, O.O., Harrison, S.T.L., van Hille, R.P., 2012. Effect of culture conditions on the competitive interaction between lactate oxidizers and fermenters in a biological sulfate reduction system. Bioresource Technology 104, 616–621.

Pattarkine, V.M.., Randall, C.W., 1999. The requirement of metal cations for EBPR removal by activated sludge. Water Science & Technology 40 (2), 159–165.

Pereira, H., Lemos, P.C.P.C., Reis, M.A.M.A.M., Crespo, J.P.S.G., Carrondo, M.J.T., Santos, H., Cresp, J.P.S.G., Carrond, M.J.T., Santos, H., 1996. Model for carbon metabolism in biological phosphorus removal processes based on in vivo13C-NMR labelling experiments. Water Research 30 (9), 2128–2138.

Peterson, S.B., Warnecke, F., Madejska, J., McMahon, K.D., Hugenholtz, P., 2008. Environmental distribution and population biology of Candidatus Accumulibacter, a primary agent of biological phosphorus removal. Environmental microbiology 10 (10), 2692–703.

Pijuan, M., Saunders, A.M., Guisasola, A., Baeza, J.A., Casas, C., Blackall, L.L., 2004. Enhanced biological phosphorus removal in a sequencing batch reactor using propionate as the sole carbon source. Biotechnology and bioengineering 85 (1), 56–67.

Pijuan, M., Ye, L., Yuan, Z., 2010. Free nitrous acid inhibition on the aerobic metabolism of poly-phosphate accumulating organisms. Water research 44 (20), 6063–72.

Poinapen, J., Ekama, G., Wentzel, M., 2009. Biological sulphate reduction with primary sewage sludge in an upflow anaerobic sludge bed (UASB) reactor–Part 4: Bed settling characteristics. Water SA.

Reichert, P., 1998. Computer Program for the Identi cation and Simulation of Aquatic Systems.

Reyes-Avila, J., Razo-Flores, E., Gomez, J., 2004. Simultaneous biological removal of nitrogen, carbon and sulfur by denitrification. Water research 38 (14–15), 3313–21.

Ribera-Guardia, A., Marques, R., Arangio, C., Carvalheira, M., Oehmen, A., Pijuan, M., 2016. Distinctive denitrifying capabilities lead to differences in N2O production by denitrifying polyphosphate accumulating organisms and denitrifying glycogen accumulating organisms. Bioresource Technology 219, 106–113.

Rickard, L.F., Mcclintock, S.A., 1992. Potassium and Magnesium Requirements for Enhanced Biological Phosphorus Removal from Wastewater. Water Science & Technology 26 (9), 2203–2206.

Rittmann, B.E., McCarty, P.L., 2001. Environmental biotechnology : principles and applications. McGraw-Hill, Boston.

Roesser, M., Müller, V., 2001. Osmoadaptation in bacteria and archaea: common principles and differences. Environmental microbiology 3 (12), 743–54.

Saad, S.A., Welles, L., Abbas, B., Lopez-vazquez, C.M., Loosdrecht, M.C.M. Van,

Brdjanovic, D., 2016. Denitrification of nitrate and nitrite by "Candidatus Accumulibacter phosphatis" clade IC. Water Research 105, 97–109.

Saad, S., Welles, L., Lopez-Vazquez, C.M., Brdjanovic, D., 2013. Sulfide Effects on the Anaerobic Kinetics of Phosphorus-Accumulating Organisms.

Satoh, H., Mino, T., Matsuo, T., 1992. Uptake of organic substrates and accumulation of polyhydroxyalkanoates linked with glycolysis of intracellular carbohydrates under anaerobic conditions in the biological excess phosphate removal processes. Water Science and Technology 26 (5–6), 933–942.

Satoh, H., Mino, T., Matsuo, T., 1994. Deterioration of enhanced biological phosphorus removal by the domination of microorganisms without polyphosphate accumulation. Water Science and Technology.

Saunders, a M., Oehmen, a, Blackall, L.L., Yuan, Z., Keller, J., 2003. The effect of GAOs (glycogen accumulating organisms) on anaerobic carbon requirements in full-scale Australian EBPR (enhanced biological phosphorus removal) plants. Water science and technology : a journal of the International Association on Water Pollution Research 47 (11), 37–43.

Schuler, A.J., 2005. Diversity matters: dynamic simulation of distributed bacterial states in suspended growth biological wastewater treatment systems. Biotechnology and bioengineering 91 (1), 62–74.

Schuler, A.J., Jenkins, D., 2003. Enhanced biological phosphorus removal from wastewater by biomass with different phosphorus contents, Part I: Experimental results and comparison with metabolic models. Water environment research : a research publication of the Water Environment Federation 75 (6), 485–98.

Schulz, H.N., Brinkhoff, T., Ferdelman, T.G., Mariné, M.H., Teske, a, Jorgensen, B.B., 1999. Dense populations of a giant sulfur bacterium in Namibian shelf sediments. Science (New York, N.Y.) 284 (1999), 493–495.

Schulz, H.N., Schulz, H.D., 2005. Large sulfur bacteria and the formation of phosphorite. Science (New York, N.Y.) 307 (5708), 416–8.

Schwedt, A., Kreutzmann, A.C., Polerecky, L., Schulz-Vogt, H.N., 2012. Sulfur respiration in a marine chemolithoautotrophic Beggiatoa strain. Frontiers in Microbiology 2 (JAN), 1–8.

Sears, K., Alleman, J.E., Barnard, J.L., Oleszkiewicz, J. a., 2004. Impacts of reduced sulfur components on active and resting ammonia oxidizers. Journal of Industrial Microbiology and Biotechnology 31, 369–378.

Seviour, R.J., Mino, T., Onuki, M., 2003a. The microbiology of biological phosphorus removal in activated sludge systems. FEMS Microbiology Reviews 27 (1), 99–127.

Seviour, R.J., Mino, T., Onuki, M., 2003b. The microbiology of biological phosphorus removal in activated sludge systems. FEMS Microbiology Reviews 27 (1), 99–127.

Seviour, R.J., Nielsen, P.H., 2010. Microbial Ecology of Activated Sludge. IWA Publishing.

Skennerton, C.T., Barr, J.J., Slater, F.R., Bond, P.L., Tyson, G.W., 2014. Expanding our view of genomic diversity in Candidatus Accumulibacter clades. Environmental microbiology 17, 1574–1585.

Sleator, R.D., Hill, C., 2002. Bacterial osmoadaptation: the role of osmolytes in bacterial stress and virulence. FEMS microbiology reviews 26 (1), 49–71.

Smolders, G.J., van der Meij, J., van Loosdrecht, M.C., Heijnen, J.J., 1994a. Stoichiometric model of the aerobic metabolism of the biological phosphorus removal process. Biotechnology and bioengineering 44, 837–848.

Smolders, G.J.F., Meij, J. Van Der, Loosdrecht, M.C.M. Van, Heijnen, J.J., 1994b. Model of the Anaerobic Metabolism of the Biological Phosphorus Removal Process: Stoichiometry and pH Influence. Biotechnology and bioengineering 43, 461–470.

Smolders, G.J.F., van der Meij, J., van Loosdrecht, M.C.M., Heijnen, J.J., 1995. A structured metabolic model for anaerobic and aerobic stoichiometry and kinetics of the biological phosphorus removal process. Biotechnology and Bioengineering 47 (3), 277–287.

Tang, S.L., Derekt, P.., Damien, C.C.K., 2007. Engineering and costs of Dual water supply system.

Tayà, C., Garlapati, V.K., Guisasola, A., Baeza, J. a, 2013. The selective role of nitrite in the PAO/GAO competition. Chemosphere 93 (4), 612–8.

Thauer, R.K., Jungermann, K., Decker, K., 1977. Energy conservation in chemotrophic anaerobic bacteria. Bacteriological reviews 41 (1), 100–180.

Tu, Y., Schuler, A.J., 2013. Low Acetate Concentrations Favor Polyphosphate-Accumulating Organisms over Glycogen-Accumulating Organisms in Enhanced Biological Phosphorus Removal from Wastewater. Enviromental Science & Technology 47, 3816–3824.

Vargas, M., Guisasola, A., Artigues, A., Casas, C., Baeza, J.A., 2011. Comparison of a nitrite-based anaerobic–anoxic EBPR system with propionate or acetate as electron donors. Process Biochemistry 46 (3), 714–720.

Wagner, M., Erhart, R., Manz, W., Amann, R., Lemmer, H., Wedi, D., Schleifer, K.H., 1994. Development of an rRNA-targeted oligonucleotide probe specific for the genus Acinetobacter and its application for in situ monitoring in activated sludge. Applied and Environmental Microbiology 60 (3), 792–800.

Wang, J., Lu, H., Chen, G.-H., Lau, G.N., Tsang, W.L., van Loosdrecht, M.C.M., 2009a. A novel sulfate reduction, autotrophic denitrification, nitrification integrated (SANI) process for saline wastewater treatment. Water research 43 (9), 2363–2372.

Wang, Q., Ye, L., Jiang, G., Jensen, P.D., Batstone, D.J., Yuan, Z., 2013. Free Nitrous Acid (FNA)-Based Pretreatment Enhances Methane Production from Waste Activated Sludge.

Environmental Science & Technology 47 (20), 11897–11904.

Wang, Y., Qian, P.Y., 2009b. Conservative fragments in bacterial 16S rRNA genes and primer design for 16S ribosomal DNA amplicons in metagenomic studies. PLoS ONE 4 (10).

Wang, Z., Meng, Y., Fan, T., Du, Y., Tang, J., Fan, S., 2014. Phosphorus removal and N2O production in anaerobic/anoxic denitrifying phosphorus removal process: long-term impact of influent phosphorus concentration. Bioresource Technology.

Wanner, J., Kucman, K., Ottová, V., Grau, P., 1987. Effect of anaerobic conditions on activated sludge filamentous bulking in laboratory systems. Water Research 21 (12), 1541–1546.

Weissbrodt, D.G., Shani, N., Holliger, C., 2014. Linking bacterial population dynamics and nutrient removal in the granular sludge biofilm ecosystem engineered for wastewater treatment. FEMS Microbiology Ecology 88 (3), 579–595.

Welles, L., Lopez-Vazquez, C.M., Hooijmans, C.M., van Loosdrecht, M.C.M., Brdjanovic, D., 2015a. Impact of salinity on the aerobic metabolism of phosphate-accumulating organisms. Applied Microbiology and Biotechnology 99 (8), 3659–3672.

Welles, L., Tian, W.D., Saad, S., Abbas, B., Lopez-Vazquez, C.M., Hooijmans, C.M., van Loosdrecht, M.C.M., Brdjanovic, D., 2015b. Accumulibacter clades Type I and II performing kinetically different glycogen-accumulating organisms metabolisms for anaerobic substrate uptake. Water Research 83 (0), 354–366.

Wentzel, M., Lötter, L., Loewenthal, R., Marais, G., 1986. Metabolic behaviour of Acinetobacter spp. in enhanced biological phosphorus removal- a biochemical model. Water SA 12 (4), 7700.

Wentzel, M.C., Ekama, G.A., Loewenthal, R.E., Dold, P.L., Marais, G., 1989. Enhanced polyphosphate organism cultures in activated sludge systems. Part II: Experimental behaviour. Water S.A. 15 (2), 71–88.

Williams, T.M., Unz, R.F., 1985. Filamentous sulfur bacteria of activated sludge: Characterization of Thiothrix, Beggiatoa, and Eikelboom type 021N strains. Applied and Environmental Microbiology 49 (4), 887–898.

Wong, M.T., Mino, T., Seviour, R.J., Onuki, M., Liu, W.T., 2005. In situ identification and characterization of the microbial community structure of full-scale enhanced biological phosphorous removal plants in Japan. Water Research 39 (13), 2901–2914.

Wong, M.T., Tan, F.M., Ng, W.J., Liu, W.T., 2004. Identification and occurrence of tetrad-forming Alphaproteobacteria in anaerobic-aerobic activated sludge processes. Microbiology 150 (11), 3741–3748.

Wu, D., Ekama, G. a., Wang, H.G., Wei, L., Lu, H., Chui, H.K., Liu, W.T., Brdjanovic, D., Van Loosdrecht, M.C.M., Chen, G.H., 2014. Simultaneous nitrogen and phosphorus removal in the sulfur cycle-associated Enhanced Biological Phosphorus Removal (EBPR) process. Water Research 49, 251–264.

Wu, G., Rodgers, M., 2010. Inhibitory effect of copper on enhanced biological phosphorus removal. Water Science & Technology 62 (7), 1464.

Xu, X., Chen, C., Wang, A., Fang, N., Yuan, Y., Ren, N., Lee, D., 2012. Enhanced elementary sulfur recovery in integrated sulfate-reducing, sulfur-producing rector under micro-aerobic condition. Bioresource technology 116, 517–21.

Xu, X., Chen, C., Wang, A., Guo, H., Yuan, Y., 2014. Kinetics of nitrate and sulfate removal using a mixed microbial culture with or without limited-oxygen fed. Applied Microbiology Biotechnology 98, 6115–6124.

Yamamoto, R., Komori, T., Matsui, S., 1991. Filamentous bulking and hindrance of phosphate removal due to sulfate reduction in activated sludge. Water Science & Technology 23, 927–935.

Yeoman, S., Stephenson, T., Lester, J., Perry, R., 1988. The removal of phosphorus during wastewater treatment: a review. Environmental Pollution 49.

Yeshi, C., Hong, K.B., van Loosdrecht, M.C.M., Daigger, G.T., Yi, P.H., Wah, Y.L., Chye, C.S., Ghani, Y.A., 2016. Mainstream partial nitration and anammox in a 200,000 m3/day activated sludge process in Singapore: scale-down by using laboratory fed-batch reactor. Water Science and Technology 74 (1), 48–56.

Zeng, R.J., van Loosdrecht, M.C.M., Yuan, Z., Keller, J., 2003a. Metabolic model for glycogen-accumulating organisms in anaerobic/aerobic activated sludge systems. Biotechnology and bioengineering 81 (1), 92–105.

Zeng, R.J., Yuan, Z., Keller, J., 2003b. Enrichment of denitrifying glycogen-accumulating organisms in anaerobic/anoxic activated sludge system. Biotechnology and bioengineering 81 (4), 397–404.

Zhao, Y., Ren, N., Wang, A., 2008. Contributions of fermentative acidogenic bacteria and sulfate-reducing bacteria to lactate degradation and sulfate reduction. Chemosphere 72 (2), 233–42.

Zhao, Y.-G., Wang, A.-J., Ren, N.-Q., 2010. Effect of carbon sources on sulfidogenic bacterial communities during the starting-up of acidogenic sulfate-reducing bioreactors. Bioresource technology 101 (9), 2952–9.

Zhou, Y., Ganda, L., Lim, M., Yuan, Z., Ng, W.J., 2012. Response of poly-phosphate accumulating organisms to free nitrous acid inhibition under anoxic and aerobic conditions. Bioresource Technology 116, 340–347.

Zhou, Z., Xing, C., An, Y., Hu, D., Qiao, W., Wang, L., 2014. Inhibitory effects of sulfide on nitrifying biomass in the anaerobic-anoxic-aerobic wastewater treatment process. Journal of Chemical Technology and Biotechnology 89 (April), 214–219.

Zilles, J.L., Peccia, J., Kim, M., Hung, C., Noguera, D.R., 2002. Involvement of Rhodocyclus -Related Organisms in Phosphorus Removal in Full-Scale Wastewater Treatment Plants Involvement of Rhodocyclus -Related Organisms in Phosphorus Removal in Full-Scale Wastewater Treatment Plants. Applied and Environmental Microbiology 68 (6), 2763–2769.

9

Annex

Effect of electron acceptors on sulphate reduction activity at WWTP

Annex 2.1

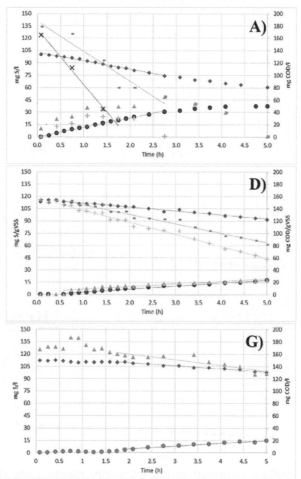

Annex 2.1.- Profiles observed in the inihibitory oxygen tests (A,D,G) showing the concentrations of sulphide (circle), sulphate (diamond), acetate (triangle), propionate (cross), lactate (X), soluble organic COD (dash) profiles in the control test performed with A) lactate, D) propionate or G) acetate as carbon source.

Annex 2.1 (cont.)

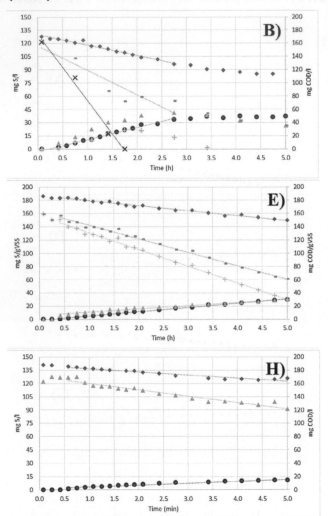

Annex 2.1 (cont.).- Profiles observed in the inihibitory nitrate tests (B,E,F) showing the concentrations of sulphide (circle), sulphate (diamond), acetate (triangle), propionate (cross), lactate (X), soluble organic COD (dash) profiles in the control test performed with B) lactate, E) propionate or F) acetate as carbon source.

Annex 2.1 (cont.)

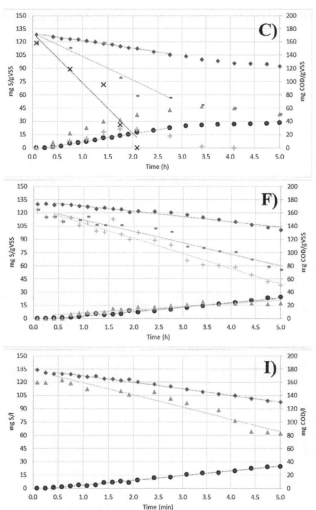

Annex 2.1 (cont.).- Profiles observed in the inihibitory nitrite tests (C,F,I) showing the concentrations of sulphide (circle), sulphate (diamond), acetate (triangle), propionate (cross), lactate (X), soluble organic COD (dash) profiles in the control test performed with C) lactate, F) propionate or I) acetate as carbon source.

Long-term effects of sulphide on the enhanced biological removal of phosphorus: The role of *Thiothrix caldifontis*

Annex 4.A

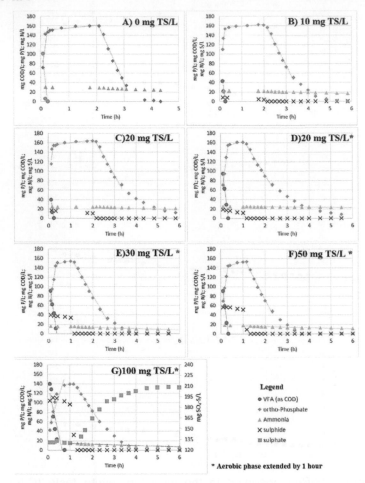

Annex 4.A.- Phosphorus (diamond), VFA as COD (circle), Ammonia (triangle), Sulphide (cross) and sulphate (square) profiles measured during the cycle test performed at 0 mg S/L (A) ,10 mg S/L (B) and 20 mgS/L (C) with an aerobic hydraulic retention time of 4 h, and at 20 mg S/L (D), 30 mgS/L (E), 50 mgS/L (F) and 100 mgS/L (G) with an aerobic hydraulic retention time of 5 h.

Annex 4.B

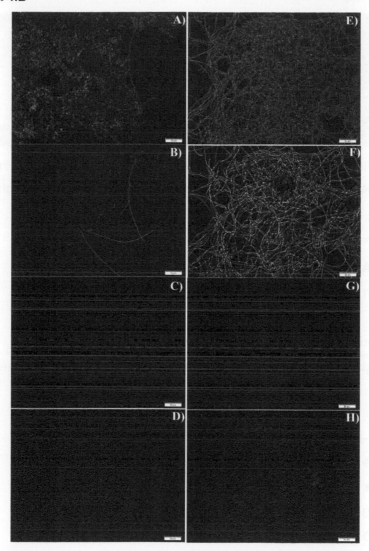

Annex 4.B- FISH analysis performed at 10 mg S/L (A-D) and at 100 mg S/L (E-H) displaying: all living organism in green (DAPI); GAO in blue (GB and DF215,618,988,1020 FISH probes); in red Candidatus Accumulibacter *phosphatis* (FISH probes PAO 462, 651, 846); and, in yellow *Thiothrix* (FISH probe G123T).

Annex 4.C

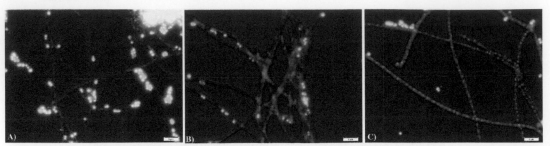

Annex 4.C.- Stained Poly-P inclusion with DAPI, in filamentous bacteria at the start of the anaerobic phase (A), end of the anaerobic phase (B) and end of the aerobic phase (C).

Annex 4.D

Mass Balance calculations

Annex 4.D1

Store carbon

Base

- Calculation based on the profiles during the cycle at 100 mg S/L
- FISH quantification was used to estimate the bio volume of Accumulibacter and *Thiothrix* (f_{FISH} ; 33% and 65%, respectively).

Symbols

$COD_s \rightarrow$ Soluble COD

$HRT \rightarrow$ Hydraulic retention time

$SRT \rightarrow$ Solids retention time

$X \rightarrow$ Biomass

$X_{PAO} \rightarrow$ Active biomass of Accumulibacter

$X_{Thio} \rightarrow$ Active biomass of Thiothrix

$Y_{sx} \rightarrow$ Observed growth

Data

VSS=1.081 gVSS/L

SRT=4.6d

HRT=16h

COD=400 mg COD/L

$Y_{sx,PAO}$=0.35 gVSS/gCOD   (Smolder et al., 1996)

Equations:

1) $X = f_{FISH} \cdot VSS$

2) $Y_{sx} = X \cdot HRT/(COD_s \cdot SRT)$ (Henze et al., 2008)

Calculations

$X_{PAO} = 0.33 \cdot 1.081 = 0.356$ gVSS/L

Based on equation 2, and in the observed growth reported by Smolders et al., (1996) for PAO is possible to calculate the COD consumption by PAO

$COD_{PAO} = 0.356*0.66d/(4.6*0.35) = 0.147$ gCOD

Conclusive remarks

The system is fed with 400 mg COD (and none leak to the aerobic phase), then the amount of XPAO cannot store all available COD. Moreover, the sulphide profile does not considerable change, which make improbable any reduction (therefore carbon consumption) or oxidation (electron source) of elemental sulfur or sulphate Thus, this suggests that other organism such as Thiothrix could store carbon as PHA.

Annex D2

Poly phosphate storage

Base

- Calculation based on the profiles during the cycle at 100 mg S/L
- FISH quantification was used to estimate the bio volume of Accumulibacter and *Thiothrix* (f_{FISH} ; 33% and 65%, respectively).
- Waste of sludge was performed during mixed conditions

Symbols

f_{pp} →   Fraction of polyP

HRT→   Hydraulic retention time

Q_{was}→   Waste sludge flow

ΔPO_4→   Phosphate removed per cycle

SRT→   Solids retention time

X→   Biomass

X_{PAO}→   Active biomass of Accumulibacter

X_{Thio}→   Active biomass of Thiothrix

Data

VSS=1.081 gVSS/L

SRT=4.6d

HRT=16h

ΔPO_4-P=37.5 mg P

$f_{pp,PAO}$=0.38 mgP/mgVSS   (Wentzel et al., 1989)

Equations

3) $\Delta PO_4 = X_{PAO,Was} \cdot f_{pp,PAO} + X_{Thio,Was} \cdot f_{pp,Thio}$

4) $Q_{was} = V/(SRT \cdot nC)$ (Henze et al., 2008)

Calculations

Based on equation 1 and 4 is possible to calculate the amount of PAO and Thiothrix which is removed from the bioreactor each cycle.

$X_{PAO,WAS}$=0.356 ·0.181= 0.064gVSS

$X_{Thio,WAS}$=0.702·0.181=0.127 gVSS

Base on the maximum fraction of Poly-P reported by Wentzel et al., (1989) and the amount of PAO removed from the reactor in each cycle is possible to calculate the amount of phosphate removed by PAO

64·0.38= 24.3 mg PO_4-P

Finally, base on equation 3 and the previous results is possible to calculate the potential fraction of poly phosphate in *Thiothrix*

$f_{pp,Thio}= (37.5-24.3)/(0.181*702)=0.10$ mg P/mgVSS

Conclusive remarks

Thus, even in the case that PAO is oversaturated with Poly-P the amount of phosphorus needed to be removed by Thiothrix is above the metabolic needs.

Annex 4.D3

Anaerobic balance of and stoichiometric of Accumulibacter and Thiothrix

Base

- All carbon has been consider as acetate
- The COD distribution was assumed as described in Annex D1
- The anaerobic stoichiometry for PAOs reported by Welles et al. (2015) was assumed to be similar for the PAO of this system

	Ratio	Unit	Measured	Estimated for Accumulibacter[a]	Estimated for Thiothrix[b]
Anaerobic	PHV/PHB	C-mol/C-mol	0.39	0.07	2.42
	PHV/VFA	C-mol/C-mol	0.21	0.09	0.29
	PHB/VFA	C-mol/C-mol	0.54	1.27	0.12
	PH₂MV/VFA	C-mol/C-mol	N.M.	N.M.	0.17[c]
	GLY/VFA	C-mol/C-mol	0.20	0.29	0.14
	P/VFA	P-mol/C-mol	0.64	0.64	0.64
Anaerobic net	PHV/L	C-mmol/L	1.34	0.20	1.14
	PHB/L	C-mmol/L	3.39	2.92	0.47
	PH₂MV/L	C-mmol/L	N.M	N.M.	0.68[c]
	P/L	P-mmol/L	4.00	1.47	2.53
	GLY/L	C-mmol/L	1.25	0.67	0.58
	VFA/L	C-mmol/L	6.25	2.3	3.95

[a] Values obtained based on Welles et al. (2015).
[b] Values calculated according to the bio-volume of *Thiothrix sp.* and assuming that acetate was the main carbon source.
[c] Calculated based on the stoichiometric conversions of the reductive branch of the TCA cycle proposed by Zeng et al. (2003)
N.M. Not measured.

Aerobic energy balance

Base

- No accumulation of glycogen nor PHA over time
- Based on the similarity of free energy between PHB and polyS (-30.8 and -19.4; McCarty, 2007; Kelly, 1999) the catabolic energy was assumed to be similar among PolyS and PHA. It was calculated based on the equation proposed by Smolders et al. (1994) of Oxidative phosphorylation and based equation proposed by Kelly (1999) of sulphur catabolism
- All other equations proposed by Smolders et al. (1994) for the aerobic metabolism of PAO, were used to calculate the energy balance.

Equations

5) $-CH_{1.5}O_{0.5}-1.5H_2O+2.25NADH_2+0.5ATP+CO_2=0$ (Smolders et al., 1994)
6) $-polyS^{2-}/S^0-4H_2O+3NADH_2+H_2SO_4=0$ (Kelly et al., 1999)
7) $-NADH_2-0.5O_2+H_2O+\delta ATP=0$ (Smolders et al., 1994)

Data
$\delta= 1.85$ (Smolders et al., 1994)

	Ratio	Units	Measured		Accumulibacter	Thiothrix caldifontis		
			PHA[a]	PHA+Poly-S		PHA[a]	PHA[b]	PHA+Poly-S
Net conversions	PHV	C-mol/L	1.34	1.34	0.2	1.14	1.14	1.14
	PHB	C-mol/L	3.39	3.39	2.92	0.47	2.8	2.8
	PO_4-P	P-mol/L	4	4	1.47	2.5	2.5	2.5
	Glycogen	C-mol/L	1.25	1.25	0.67	0.58	0.58	0.58
	Biomass Growth	C-mol/L	3	3	0.99	1.99	1.99	1.99
	Sulphide oxidation	S-mol/L	N.A	2.43	N.A.	N.A	N.A	2.43
Energy balance of the aerobic metabolism	**Source/Use**				**NADH balance**			
	PHA		-[0.74][c]	-[0.74][c]	2.18	-[1.69][c]	1.44	1.44
	Biomass growth		1.84	1.84	0.61	1.22	1.22	1.22
	Phosphate transport		-0.44	-0.44	-0.16	-0.27	-0.27	-0.27
	Glycogen		1.25	1.25	0.67	0.58	0.58	0.58
	Sulphide oxidation		N.A	7.29	N.A	N.A	N.A.	7.29
	Balance		**2.69**	**9.94**	**3.3**	**1.53**	**2.97**	**10.26**
	Source/Use				**ATP balance**			
	Oxidative phosphorylation		4.98	18.38	6.10	2.83	5.49	18.98
	PHA		N.A	N.A	0.48	N.A	0.32	0.32
	Biomass growth		-4.50	-4.50	-1.48	-2.98	-2.98	-2.98
	Phosphate transport		-4.00	-4.00	-1.47	-2.50	-2.50	-2.50
	Glycogen		-1.04	-1.04	-0.82	-0.48	-0.48	-0.48
	Balance		**-[4.56][d]**	**8.84**	**2.81**	**-[3.13][d]**	**-[0.15][d]**	**13.34**

[a] Estimations based on the measured PHA concentrations.
[b] Estimations assuming a ratio of 1 C-mol PHA stored per 1 C-mol of VFA consumed.
[c] Insufficient carbon conditions for biomass growth and glycogen formation.
[d] Energy deficient conditions.

Cooperation between Competibacter sp. and Accumulibacter in denitrification and phosphate removal processes

Annex 5.A

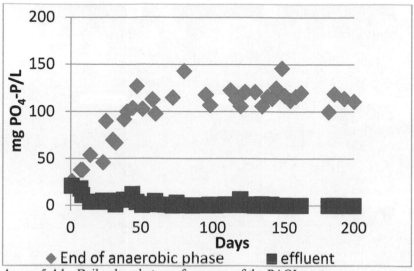

Annex 5.A1.- Daily phosphate performance of the PAOI reactor

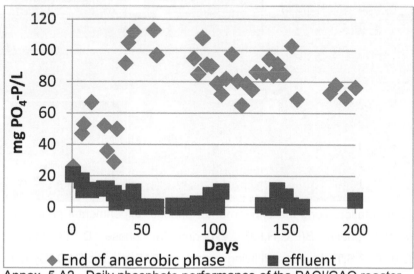

Annex 5.A2.- Daily phosphate performance of the PAOI/GAO reactor

Absent anoxic activity of PAO I on nitrate under different long-term operational conditions
Annex 6.A

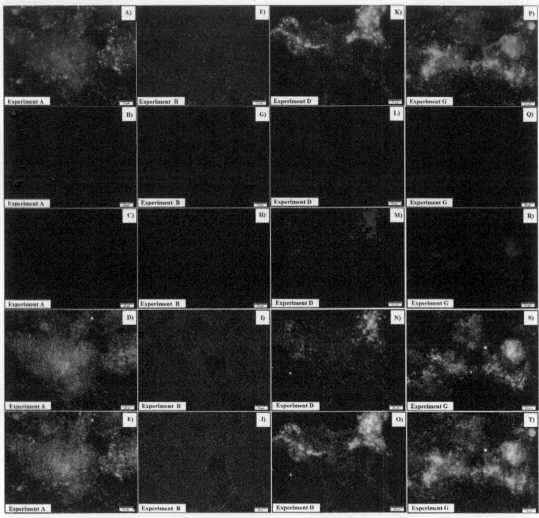

Annex 6.A.- Microbial adaptation along the experimental phases: experimental phase A (A,B,C,D,E), experimental phase B(F,G,H,I,J), experimental phase D (K,I,M,N,O) experimental phase G (P,Q,R,S,T). In green all living organism (DAPI), in red PAO (Cy3), in yellow PAOI (FAM), in blue GAO (Cy5).

About the author

 Francisco Rubio was born in 1986 in Durango, México. With a background in civil engineering, Francisco have gained work experience in different companies and projects throughout México and abroad. Nevertheless, feeling the urgency to preserve our earth and its resources and the desire to contribute, Francisco decided to obtain further knowledge and education in the field of Waste Water treatment. In October 2011 Francisco started his MSc in Municipal Water and Infrastructure, specialization Sanitary Engineering at UNESCO-IHE. During the first year of study he gained knowledge in the biological and chemical removal of nutrients, modelling of microbial process, and in the different configuration and operational conditions of wastewater treatment plants. Accordingly, in April 2013 Francisco graduated with distinction. During the research phase, he focused on the recovery of nitrogen and phosphorous (as Struvite) from waste streams, using affordable sources of magnesium such as Seawater.

Once his MSc studies concluded, he decided to continue his research through a PhD study. During his PhD research, Francisco reinforced skills such as: organization, time management, creative thinking among others. As you are able to read in this thesis his PhD research focus was mainly in two themes: i) The interaction of sulphate reduction with the biological removal of phosphorous, and ii) The phosphorous removal using nitrate as electron acceptor. In practice, both of these research themes could help to: i) Reduce sludge production, and ii) Carbon requirements of a wastewater treatment plant, while complying with the E.U. standards for the discharge of treated wastewater.

Acknowledgements

There is a mix of feelings to explain how one feels when this chapter of my life than start on October 2013 is close to finish. Part of me to be honest feels relief but there is part of me that is going to miss all those moments that I share with the people that help me behind the curtains during this journey. Now I would like to take a moment to thank all those people that help me along my PhD research.

I would like to thank first two wonderful people and professors that guide me through this study Prof. Damir Brdjanovic and Prof. Mark van Loosdrecht. Damir you help me to keep my feet on the ground and always to think in the implications of my studies. One of your phrases that I now take with me is "think out of the box", a simple phrase that can be interpreted into several meanings. Thank you Damir for always having your door open for me and for supporting me with each thing that I need during my PhD research. Mark I am really grateful for all your help, your comments and ideas help me to go into detail on the process and ask myself each time why certain process was or not happening. You help me to go more into detail, to critical question my findings and found a ground to defend them. Thank you very much to both of you Damir and Mark, because you two teach me to keep my mind open for new ideas but to do not lose focus on the main point.

Next I would like to thank two amazing scientist that today I am proud to call them friends, Carlos and Laurens. Carlos and Laurens, thank you very much for the discussion, for the brainstorming, for each coffee that helped me along this process. Carlos I want to say thanks for everything, you help me in each step during this research from the laborious writing of 3 PhD proposals (☺) until this time. Your guidance and advices helped me to grow as a scientist. Laurens I want to thank you for everything, especially at the start of my PhD to help me to grasp the scientific knowledge that I did not have. Sorry for all those accidents that happened in the lab, and thank you for helping me fix them. Carlos and Laurens, I really do not found word to express my gratitude to you guys. I will found myself lucky if in the future I can scientifically collaborate with both of you, and please keep in touch.

I also would like to thanks to the person that help me to take this road, Roberto, all of this start at your coffee table, thank you for guiding me in this direction. I would like to thank all the lab staff from UNESCO-IHE which helped me with the different analytical procedure of this research. Fred, thank you very much for all your help, in particular for facilitate the measurement of substance that were not possible to be measure at IHE. Ferdi, you were my lab support during my MSc but your help did not stop there, thanks for all your help, especially with the I.C.P. Berend, thank you for the help with the I.C. and for the assistance during the flooding times (☺). Frank, thank you for the help with the G.C. Lyzette, thank you for the help with the I.C, and for buying everything that I need for my research. Peter thank you for the help with autoclaves and microscope . But especially guys, thank you very much for all the laughs and good moment during my work in the lab.

In addition, I will like to thank to the lab staff from TU Delft. Ben, thank you for all your work and efforts in the 16s rRNA, ppk and DGGE analysis that you perform for me. Mitchell, thank you for your help with the ppk and DGGE analysis. Both of your work definitely improved my research. Dimitri, thank you very much for finding a space in your agenda and sharing your expertise on sulphide oxidizing bacteria with us.

I would like to specially thank to all the collaborators of this research. Prof. Kuenen I am grateful that you open your house to us and share with us your expertise on the sulphur cycle. Prof. Nielsen your help and comments on chapter four, help us to proof the concept of anaerobic VFA storage of *Thiothrix caldifontis*. Martha, I am really glad for all your efforts on the FISH MAR of chapter 4 I hope we can collaborate again in the near future. David I am grateful for your work and comments on chapter six, which help us to come closer to an understanding of the involvement of PAO on the anoxic dephosphatation observed in WWTP. Tessa, thank you very much for sharing your experience and knowledge on sulphate reduction. Prof. Jules I am grateful for the hallway talks even without you notice (or maybe yes) you give me some ideas on the sulphur cycle.

Even if academic, this research is not based on pure science there was some many people than helped me in different matters to finish this thesis. I am grateful to Mexican Council for Sciences and Technology (CONACYT) for the economic support given that allow me to follow my PhD research. Jolanda, thank you for your help with the budget and insurance. Jaap and Gordon, men thanks for taking care of my laptop without it I could had not write anything at all. Sylvia and Maria thank you for making me smile when I passed by in the office.

I would also like to thank to my MSc students that help me to distract my mind from my research topic, and for trusting me as mentor, Mona, Alvan, Mosses and Dykabelo, thanks.

Dineke, and Enrique (Hendrikus ☺) thank you very much for helping us with more than just my PhD. To all my extent family, thanks for all the good moments lets hope there is more to come. Opa and Oma thank you very much for opening your house and heart to me.

Gordo y mamá muchas gracias por todo, ustedes me pusieron en este camino desde chiquito. Sin ustedes jamás hubiera llegado a donde estoy. Su amor y suporte me ayudaron a superar cada obstáculo y a pesar de la distancia los sentí cerca de mi a cada paso. Gloria y Karla , mis hermanitas las quiero mucho cada una de ustedes me ayudo a mejorar a su manera y a su momento. Diego, Carlos y Mariana aunque no se los diga los quiero y su felicidad y alegría aligero mis estudios. Tata y Mami Alicia, su ejemplo y amor supera las palabras gracias por estar conmigo. Y a toda mi familia Rubio y Rincon por sus palabras de aliento y consejos, gracias.

At last, the most important person that help me with this research, that grab my hand and never let it go and support me in each step, or better say that each step we took together Marielle this thesis is yours as well. I am grateful for finding a person which I feel so complete and so happy with, Marielle my love you bring happiness to my life. You give me the force and energy to keep going, you help me to smile on the darkest days, you my dear give me more than inspiration you give me a reason to be the man that I am today. I am grateful for you little one, in this same moment you and your mom are helping me to keep going. You had not come to this world yet but there is some much love and happiness around that is fulling everyone, and all because of you thanks for your gift, thanks for you.

Netherlands Research School for the
Socio-Economic and Natural Sciences of the Environment

D I P L O M A

For specialised PhD training

The Netherlands Research School for the
Socio-Economic and Natural Sciences of the Environment
(SENSE) declares that

Francisco Javier Rubio Rincón

born on 29 January 1986 in Durango, Mexico

has successfully fulfilled all requirements of the
Educational Programme of SENSE.

Delft, 30 January 2017

the Chairman of the SENSE board

Prof. dr. Huub Rijnaarts

the SENSE Director of Education

Dr. Ad van Dommelen

The SENSE Research School has been accredited by the Royal Netherlands Academy of Arts and Sciences (KNAW)

KONINKLIJKE NEDERLANDSE
AKADEMIE VAN WETENSCHAPPEN

The SENSE Research School declares that Mr Francisco Rubio Rincón has successfully
fulfilled all requirements of the Educational PhD Programme of SENSE with a
work load of 30.1 EC, including the following activities:

SENSE PhD Courses

o SENSE writing week (2014)
o Environmental research in context (2014)
o Research in context activity: 'Co-organizing the UNESCO-IHE PhD Symposium on:
   Integrating research water sectors' (2015)

Other PhD and Advanced MSc Courses

o Anaerobic wastewater treatment, UNESCO-IHE (2015)
o Online course Biological wastewater treatment, UNESCO-IHE (2015)
o Online course Modelling sanitation systems, UNESCO-IHE (2015)

Management and Didactic Skills Training

o Supervising three MSc theses (2014-2016)
o Assisting laboratory classes of the MSc 'Sanitary engineering' for the course 'Biological
   wastewater treatment: Principles, modelling, and design' (2014-2016)
o Reviewer for the journals 'Water Research' and 'Chemosphere' (2015-2016)
o Supervising students in the MSc 'Sanitary engineering' during group work for the course
   'Case study of the wastewater treatment plant at Sint Maarten' (2015-2016)

Oral Presentation

o *Aerobic sulphide effects on EBPR.* Water Environment Federation (WEF/IWA) Nutrient
   Removal and Recovery Conference, 10-13 July 2016, Denver (CO), USA

SENSE Coordinator PhD Education

Dr. ing. Monique Gulickx